# Practical Manual on Plantation Forestry

**Dr. Pankaj Panwar**

B.Sc. (For.), M.Sc. (For.), Ph.D. (For.)
Assistant Professor,
Department of Forestry, Uttar Banga Krishi Viswavidyalaya,
Pundibari, Cooch Behar - 736 165 (W.B.)

**Dr. S.D. Bhardwaj**

B.Sc. (Agri.), M.Sc. (Agri.), Ph.D. (Agri.), PDF (Utah)
Sr. Scientist,
Department of Silviculture and Agroforestry,
Dr. Y.S. Parmar University of Horticulture and Forestry,
Nauni, Solan - 173 230 (H.P.)

SCIENTIFIC PUBLISHERS

*Published by:*

Scientific Publishers (India)
5 A, New Pali Road, P.O. Box 91
Jodhpur 342 001 (India)

*Branch Office*
Scientific Publishers (India)
4806/24, Ansari Road,
Daryaganj
New Delhi - 110 002 (India)

E-mail: info@scientificpub.com
Website: www.scientificpub.com

Print: 2016

ISBN: 978-81-7233-429-1



Lasertype set : Rajesh Ojha

Printed in India

*Dr. S.S. Negi* IFS
Director

**FOREST RESEARCH INSTITUTE**
P.O. NEW FOREST
DEHRA DUN - 248 006

# Foreword

Forests, not only provide multiple benefits to mankind but, also help in conserving the environment and maintaining the ecological balance. In fact, they provide vital ecological security to a region or nation. With this, the role of forest plantations has become very important. It had been estimated that, global rate of new plantation is 4.5 million hectares per year, of which, 79 per cent is in Asia. India stands among the five nations which accounts for 65 per cent of world's forestry plantations. Good quality, fast growth, cheap and efficient harvesting, lower transport and processing cost of plantations are some of the advantages of plantation forestry. These plantation forests are a viable option for decreasing the ever increasing gap between demand and supply of fuel and timber wood in India.

Realizing the importance of "Plantation Forestry", it was made a compulsory subject at the undergraduate level and optional subject at the postgraduate level of forestry, to produce technical manpower well versed with plantation techniques. The book entitled "Practical Manual on Plantation

Forestry" by Dr. Pankaj Panwar and Dr. S.D. Bhardwaj is a noble effort, which will fulfil the gap in the documented literature on this subject.

I am confident, that, this book will be widely useful for practical foresters, farmers and academicians. This will be particularly helpful to students pursuing their career in forestry, as it covers all the topics of the "Plantation Forestry" course of the undergraduate level.

**DEHRA DUN**

22.06.2005 **Dr. S.S. Negi,** IFS

# Preface

In wake of ever increasing gap between demand and supply of fuel wood, timber and other wood based products and also concern of growing pollution due to burning of fossil fuel, the need is to increase the forest cover by way of planting trees artificially. This necessitates training and imparting knowledge of plantation and its related aspects to all concerns who want to make forestry as their career. Plantation forestry is one of the compulsory course to all students pursuing their B.Sc. (Forestry). In the absence of a uniform course throughout India, the course content of plantation forestry differs from one university to another. In this manual attempt has been made to cover important topics of plantation forestry syllabus of different universities. The course content of plantation forestry of Dr. Y.S. Parmar University of Horticulture and Forestry, Solan (Himachal Pradesh), Tamil Nadu Agriculture University, Coimbatore and Indian Council of Agriculture Research (ICAR), New Delhi is given in appendix. The manual contains details on tools and equipments which are used in forestry operations, techniques of planting trees, methods of tending forest crops, points to be kept in mind while planning of plantation, impact of plantations on biodiversity, soil, society and economics, different sources of energy, energy potential of biomass and plant protection and economics. In each topic, efforts have been given towards practical aspects and step-wise-step execution of the work. Few questions have also been given at the end of each chapter to examine the extent to which the subject was understood by the reader.

The authors hope, that, this practical manual will be useful for imparting practical skills to the students and practical foresters. Suggestions to improve this manual are welcome.

**Dr. Pankaj Panwar**
**Dr. S.D. Bhardwaj**

# Contents

# Chapter 1

# Introduction

The practice of planting trees goes back to ancient times, and many economically important species have been widely planted outside their natural range for thousands of years. Prior to 1900, low population density and the widespread availability of natural forests meant that there was no need to plant trees extensively as an industrial resource. However, with increase in population and also increase in per capita consumption of wood products viz. paper, furniture, timber etc, a need was felt to introduce some fast growing species to bridge the gap between demand and supply of wood for industrial purposes. *Poplars* and *Eucalyptus* were extensively grown because of their fast growth and wide adaptability. These fast growing large scale forest plantation were termed as "Plantation Forests". These are defined as "trees established on a sufficient scale, and with income generation ranking sufficiently high as one of the purposes, to have reasonable potential as a commercial crop". This definition includes various physical configurations and production systems, not just even-aged monoculture blocks grown for wood production, and therefore various forms of farm (agro-) and community forestry, but excludes urban forests and rural plantings without commercial intent. In the first half of the 20th century tree planting began in earnest in Western Europe, the United States, Australia, New Zealand, South Africa and a small number of developing countries such as India, Chile, Indonesia and Brazil. Later, in the 1950s, Japan, Korea and China embarked on massive reforestation programmes (Evans, 1992).

In 1960s large-scale plantation programmes were launched in many tropical and subtropical countries, and between 1965 and 1980 the area devoted to tropical plantations increased tremendously. During this period, the United Nations Food and Agriculture Organisation (FAO) played an important role by disseminating technical information and promoting plantations. In most cases, plantations were established with financial support from foreign donors or with soft loans. According to the *Global Forest Resource Assessment 2002*, conducted by FAO, the global plantation estate increased from 17.8 million hectares in 1980 to 43.6 million hectares in 1990 and 187 million hectares in 2000 (FAO, 2001a). A third of today's plantations are found in the tropics and two thirds in temperate and boreal zones. A relatively small number of countries dominates the plantation business, with five, each possessing over 10 million hectares of plantations, accounting for 65 per cent of the world's plantations. These are China, the United States, the Russian Federation, India and Japan. However, few of their plantations could be classified as fast wood. The FAO assessment estimates the global rate of new planting at 4.5 million hectares a year, with Asia accounting for 79 per cent and South America 11 per cent (Cossalter and Smith, 2003). The reason for the rapid expansion of large-scale plantations is purely economic. The fast growing plantations can produce one and a half to two times more wood per hectare per year and reach maturity two to three times faster, than longer-rotation softwood plantations. Good-quality fast wood comes from plantations where there is uniformity in tree size and shape. This leads to cheap and efficient harvesting, and lower transport and processing costs. If the end product is to be of high quality, then there must be uniformity of certain characteristics : less internal tension for sawn timber, better opacity for pulp, and so forth. With their cheap land, low labour costs and potential for higher tree growth rates, developing countries located in the tropics and subtropics, like India, have a competitive advantage over cooler, temperate regions when it comes to producing plantation wood. Plantations come in many shapes and guises, and are established for a variety of reasons. Some provide shelter, shade and fodder for livestock; others fuelwood for households, and timber for furniture and the construction industry. Sometimes they are established for the benefit of wildlife or as a recreational resource. Plantations may even provide a valuable service to urban populations, particularly in arid zones, by absorbing storm and sewage water. Plantations frequently fulfil a whole range of roles, for example — by providing peasant farmers with fodder, villagers with fuel wood and industry with high-quality timber. However, large-scale plantations have to face criticisms from environmentalists on account of their temporary nature and monocultures.

Though, 30-40 years had passed since the start of plantation forestry however, there is no standard classification of plantation forestry. *Global Forest Resource Assessment 2002,* however, recognizes three broad categories of plantations:

1. **Industrial plantations:** which produce wood or fibre to supply raw materials to wood processing industries and charcoal for industrial use;
2. **Non-industrial plantations:** which produce fuel wood for domestic use, or are established to protect soil and water resources; and

3. Plantations whose purpose and end products are unspecified.

However, FAO's figures make no distinction between fast-wood and other types of industrial plantation.

Plantation forestry though is in its infancy stage, but it has been seen as a viable option for fulfilling the ever increasing gap between demand and supply of the wood for various purposes. In the following chapters, we will be discussing one by one the pros and cons of the plantations right from establishing of the plantation to their harvesting along with the social, economical and environmental issues related to plantation forestry.

Chapter 2

# Equipments and Tools used in Plantation Forestry

In India, to practice forestry, degraded and inaccessible areas are allocated, as fertile lands are to be kept for agriculture and horticulture. These degraded and undulating lands have characteristics like shallow soil depth, less fertility, more slope (in hilly tracts), variation in aspects etc, which necessitates adopting of special techniques of preparing soil for plantation and their management. The tools which are used for preparing site, harvesting, transportation, pruning etc in forestry are presented in this section

## Equipments and materials for establishment and management of plantations

Owing to variation in land characteristics viz. deep and soft soil or hard and shallow soil or salt pans, hills and plains etc, the choice of equipments used for preparing sites differs. Also for harvesting of crops, extinguishing fire, pruning of crops, transportation of logs from forest different equipments are required.

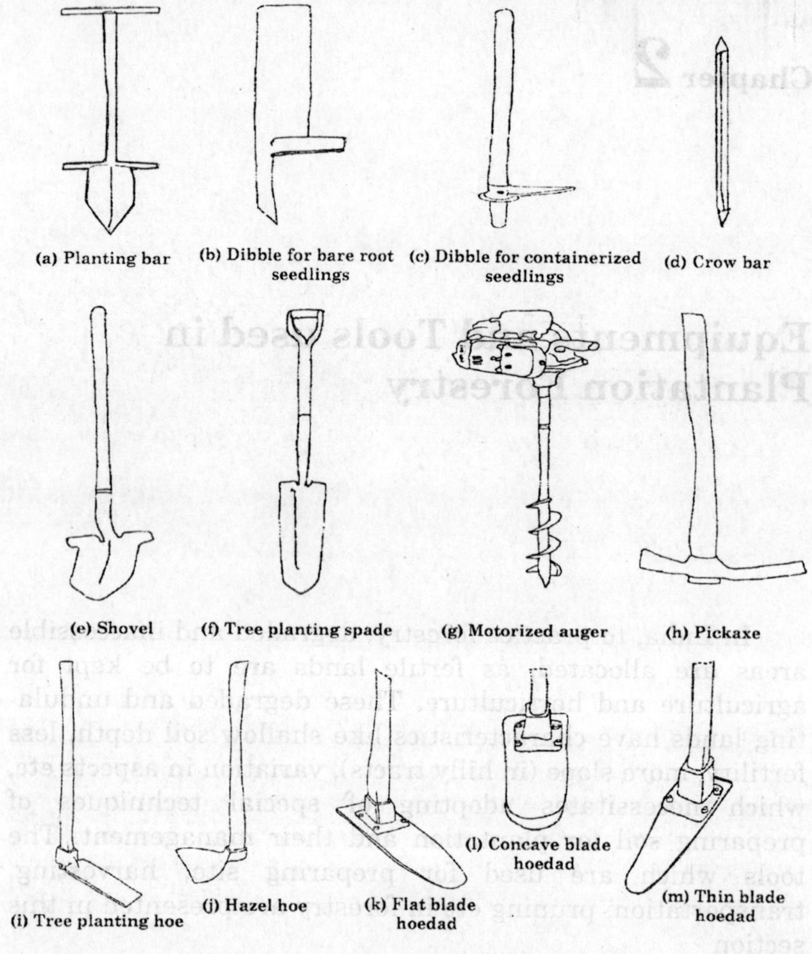


**Figure 1.** Tools used for tree planting

## (a) Hand tools for tree planting

Fig. 1 (a-m) shows the tools used for preparing site for planting of trees. Figs. 1 (a-d) are planting bars, which are used for making deep small holes in which plants/seedlings are planted. Tools in Fig. 1 (e & f) are used for soils which are light textured/ soft. Fig. 1 (g & h) are used for areas which are rocky and requires breaking of hard pans. Fig. 1 (i-m) are variation of hoes which are used in areas where soils are intermediate i.e. neither too light nor too rocky.

### (b) Tools used for controlling forest fire

Fig 2 (a-h) shows tools used for preparing fire lines to check extension of forest fire and to extinguish forest fire. Fire rake and shovel are used to remove the plant parts and leaves from the area to check fire extension from one place to other. Swatter is used to beat the fire to extinguish it and to spread soil. Mattock and pickaxe (Fig. 1h) are used to dig the soil, either to make trenches or to expose soil to reduce burning material. sickle, *Khunkhri*, axe, Pulaski are used to cut and remove the shrubs and trees while preparing fire lines. Fire extinguishers which usually contains compressed $CO_2$ are used to check fire by checking the supply of oxygen to the burning materials.

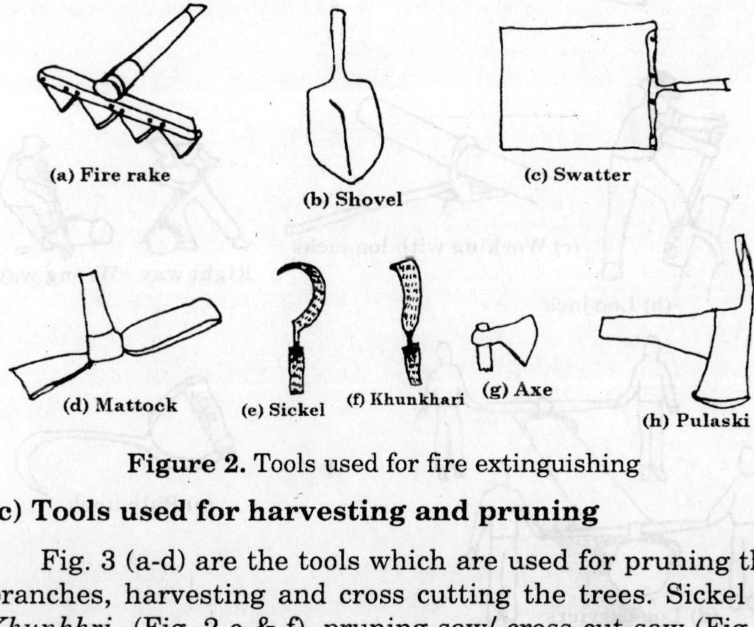


**Figure 2.** Tools used for fire extinguishing

### (c) Tools used for harvesting and pruning

Fig. 3 (a-d) are the tools which are used for pruning the branches, harvesting and cross cutting the trees. Sickel & *Khunkhri* (Fig. 2 e & f), pruning saw/ cross cut saw (Fig. 3 c), scateur are used for pruning the branches of the trees.

Axe (Fig. 2g) & saw (Fig. 3a) and power chain saw (Fig. 3d) are used for felling of the trees and cross cutting them.

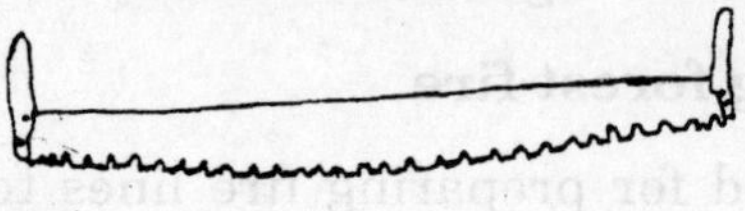

**(a) Saw**

**(b) Framed saw**

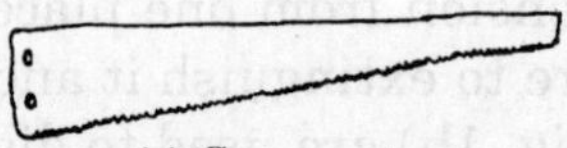

**(c) Cross cut saw**

**(d) Power chain saw**

**Figure 3.** Tools used for harvesting and pruning

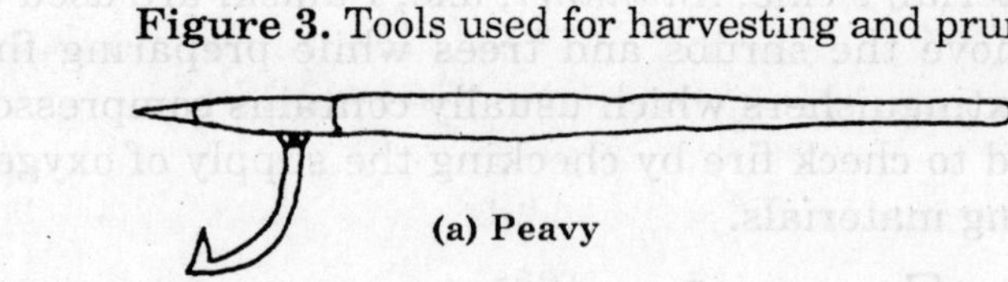

**(a) Peavy**

**(b) Log jack**

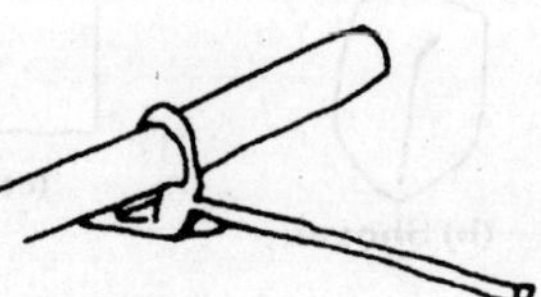

**(c) Working with log jacks**

**Right way Wrong way**

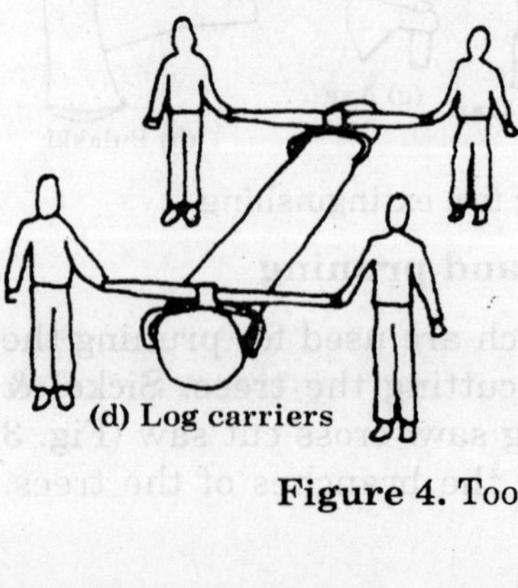

**(d) Log carriers**

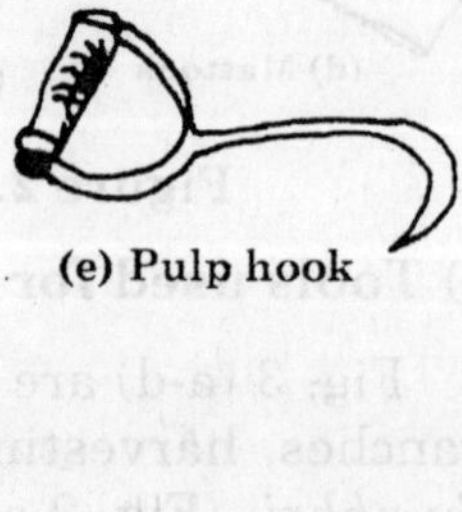

**(e) Pulp hook**

**Figure 4.** Tools for handling logs

## (d) Tools used for handling logs

Fig. 4 (a-e) are the tools used for handling logs. Peavy is used for rolling of logs during bucking and loading. Log jacks makes bucking of log easier. Method of working with log jacks is given in Fig. 3c. Log carriers are used to transport the big logs manually from one place to other. Pulp hooks are used to pull or push the small logs while loading or unloading or during stalking the small logs.

## Tools/equipments/machineries used for transportation of logs

Transportation of logs are done by two ways : either on land or over head, choice of which depends upon area and resource availability. Logs are heavy and voluminous, hence to transport them using manual labour is costly as well as time consuming. Hence in areas where elephants are available, they are used to drag the logs out of the forest or to the road head. In areas where cart roads are available, carts are also employed for transporting the logs to log yards. However, when time in hand is less tractors are employed after preparing a "kucha" road. Fig. 5 (a-e) shows the methods of transporting the logs from the forest by employing tractors. Either skidding pans or tractors with hydraulic lift or winches are employed for dragging the logs. The logs are tied with skidding tongs or with choker and are fitted to tractors with iron chains. In cases where roads are available transportation through trucks is the most preferred way.

Overhead transportation of logs is done in areas where natural gradient is available i.e in hilly terrain. Overhead transportation includes ropeways, which may be gravity ropeway or power ropeways or through helicopters. However, the latter are seldom used in India owing to their high cost.

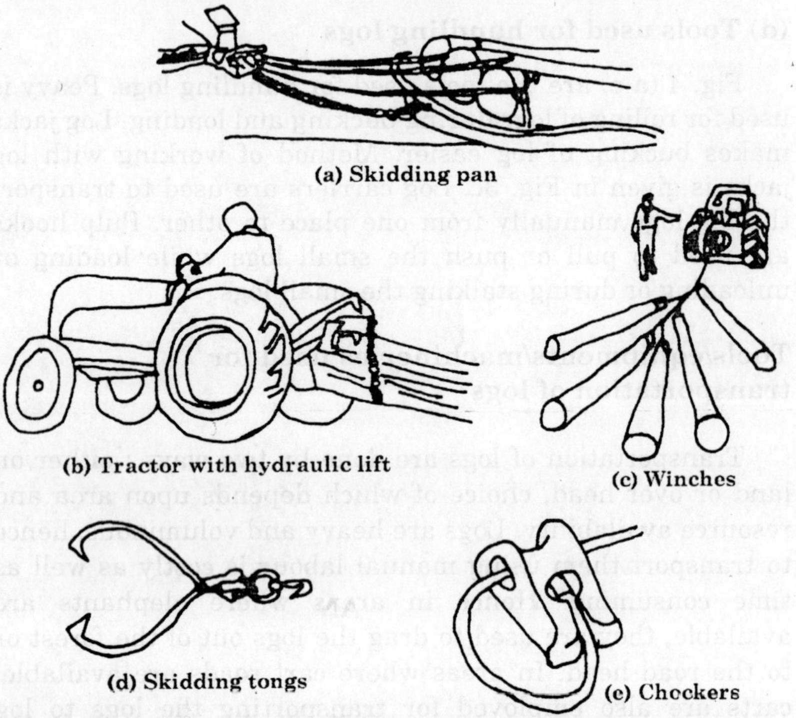

(a) Skidding pan

(b) Tractor with hydraulic lift

(c) Winches

(d) Skidding tongs

(e) Chockers

**Figure 5.** Equipments used for transporting logs within forest

## Do it yourself

1. Enlist the tools used for harvesting of trees in your area.
2. Study the variation in transportation of logs for shorter distance and longer distance in your locality
3. Use power chain saw and manual saw to cross cut a log. Find out the time saved while using power chain saw.
4. Enlist the tools used for handling logs in your locality.

Chapter **3**

# Plantation Technology

## CHOICE OF SITE AND SPECIES

To obtain expected return from the piece of land on which forestry is to be practiced, careful matching of site with species is of utmost importance. As mentioned earlier, for practicing forestry degraded and undulating areas are spared, hence, there is less liberty of choosing the site for a forester. Because of restricted choice of sites for practicing forestry, choice of species, which suits to the site and owner's objective becomes more important. Following is a tabular presentation of choice of species depending upon different climatic and edaphic conditions prevailing at the allocated site.

| Site/ situation | Species recommended |
|---|---|
| **(A) Edaphic** | |
| Ravine lands | In addition to grasses, species which spread quickly should be planted *viz. Albizzia* spp., *Leucaena leucocephala, Glyricidia maculata, Dendrocalamus strictus, Moringa oleifera* |

| | |
|---|---|
| Salt tolerant | *Tamarix aphylla, Salvadora persica, Atriplex* spp. |
| Saline and Alkaline soil | *Albizzia* spp., *Pongania pinnata, Acacia arabica, Prosopis juliflora, Azadirachta indica, Terminalia arjuna* |
| Shifting sand dunes | *Calotropis procera, Tecomella undulata, Acacia tortilis, Prosopis juliflora,Prosopis cineraria, Azadirachta indica, Acacia senegal* |
| Cold desert | *Hippophae rhamnoides, Populus alba, Populus ciliata, Salix fragilis, Salix alba* |
| Lateritic soil | *Acacia auriculiformis, Shorea robusta, Dendrocalamus strictus, Maduca latifolia, Dalbergia* spp., *Pterocarpus marsupium, Gmelina arborea.* |
| Water logged soil | *Terminalia arjuna, Eucalyptus robusta, Anthocephalus chinensis, Bombax ceiba, Salix* spp., *Syzygium cumini, Trewia nudiflora.* |
| **(B) Climate** | |
| Wet temperate region | *Salix alba, Quercus* spp., *Populus ciliata, Betula utilis, Picea, Abies* |
| Dry temperate region | *Juglans regia, Juniperus communis, Pinus gerardiana* |
| Wet subtropical region | *Albizzia* spp., *Acrocarpus fraxinifolius, Bombax ceiba, Morus* spp., *Populus* spp., *Terminalia* spp., *Grewia* spp. |
| Dry subtropical region | *Acacia catechu, Azadirachta indica, Eucalyptus* spp., *Pinus roxburghii, Dalbergia sissoo, Embilica officinalis* |
| Wet tropical region | *Michelia champaca, Dipterocarpus* spp., *Bamboos, Gmelina arborea, Sesbania grandiflora, Shorea* spp. |
| Dry tropical region | *Acacia nilotica, Diospyrous melanoxylon, Pongamia pinnata, Tectona grandis, Dendrocalamus* spp., *Cassia fistula.* |

## LAYOUT OF PLANTATIONS IN HILLS AND PLAINS

### (a) Layout of plantation in hilly area

In hills, plantation is done using contour techniques. Marking of contour line can be done either using wooden frames or Abney's level.

### Construction of the contour lines

Contour line is a imaginary line, which joins two points on the earth surface having same height above mean sea level. Contour interval is the distance between two contour lines. Contouring in forestry is needed for road alignment, soil conservation methods, planning afforestation in the hilly terrain by making contour trenches, layout for irrigation channels in forest nurseries.

### Constructing contour lines by Wooden frames

Wooden frames used for preparing contour line are of two types : one that uses plumb bob and other that uses spirit leveller.

Wooden frame using plumb bob is shown in Fig. 6a. It is in form of English alphabet "A". Its height is usually 1.5 m and distance between two legs is 2 m. End "X" and "Y" are said to be on same level when plumb bob hanging from Z point is exactly in the middle of PQ. While working with "A" type wooden frame point "X" is kept at the place from where contour line has to be started. The point "Y" is moved up and down the hill till the plumb bob rests at point R. At this situation "X" and "Y" point are on the same level. To proceed further wooden frame is rotated i.e. point "Y" is kept fixed and point "X" is turned to next place (Fig. 7). In similar way, points falling on same line are marked and contour line is drawn (adopted from Panwar and Bhardwaj, 2005).

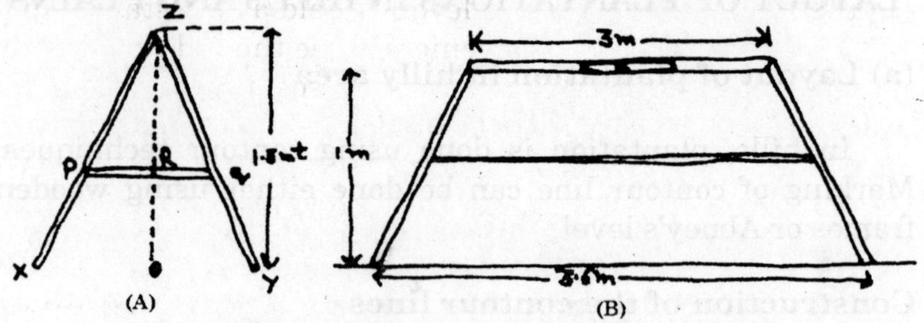


**Figure 6.** Wooden frames for marking of contour lines (Adopted from Panwar and Bhardwaj, 2005)

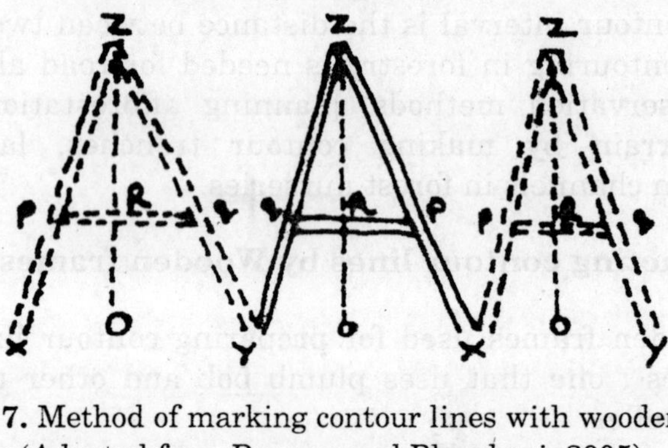


**Figure 7.** Method of marking contour lines with wooden frame (Adopted from Panwar and Bhardwaj, 2005)

Wooden frame using spirit level is shown in Fig. 6b. The spirit is fixed in the middle of top frame. The leg of the wooden frame is kept at the starting point "X". The other leg of the frame is moved up and down till the air bubble of spirit level is in middle. At this point "X" and "Y" are at same level. The frame is rotated by keeping point "Y" fixed as explained in Fig. 7.

## Procedure for preparing contour line with abney's level

The observer stands at a convenient point from where contour line needs to be drawn. In hilly areas marking of the contour lines should usually start from the foothills. A staff/ iron rod marked with black and white paint or a ribbon tied on the rod, at observers eye level, is held at a distance up to which the observer can conveniently see the rod through the abneys level. The observer tries to see the mark on the staff/rod and coincide it with the bubble and horizontal wire of abneys level without moving the abney's level. Observer instructs the man standing with the rod/ staff to move uphill or down hill so that all the three points coincide. When all the three points air bubble, horizontal wire and mark on the staff coincides, it means that the two points are on the same elevation above mean sea level. The observer then moves to the point where staff was kept and staff is taken to next place (as shown by arrow mark in Fig. 8). In this way point A and C are on the same elevation and thus makes a contour line (adopted from Panwar and Bhardwaj, 2005).

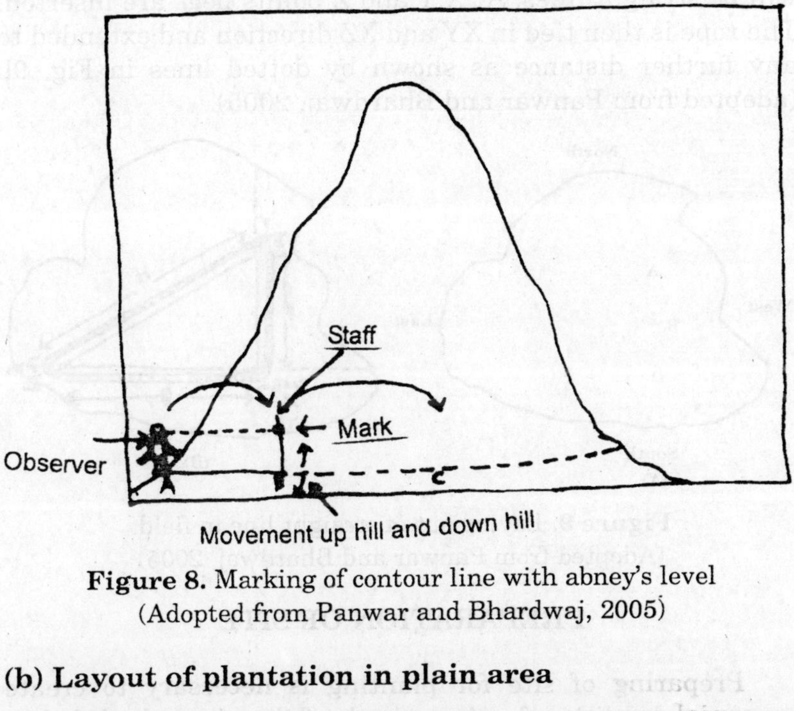


**Figure 8.** Marking of contour line with abney's level (Adopted from Panwar and Bhardwaj, 2005)

**(b) Layout of plantation in plain area**

In plains while going for layout first and foremost need is drawing a straight line, which will act as a base line for further plantations. In field it is difficult to know in which direction the line is straight. To draw a straight line we use the basis of Pythogorous theoreum i.e $H^2 = P^2 + B^2$ where H= hypotenuse, P= perpendicular and B=base). In such condition perpendicular and base will be straight line. To draw straight line in field three men are required. We take a tape and it is held at zero depicted by point "X" in Fig. 9. Now from "X" 3 m is measured and another men hold the tape at point "Y". The tape is further extended and held by third men at 7 m mark of tape depicted by point "Z". The same tape is extended and given to the men standing at point "X" where he will hold the tape at 12 m mark on the tape. All men will tighten the tape by moving backward. At this situation ∠YXZ will be $90^0$ and hence line XY and XZ will be straight lines. At X,Y and Z points pegs are inserted. The rope is then tied in XY and XZ direction and extended to any further distance as shown by dotted lines in Fig. 9b (adopted from Panwar and Bhardwaj, 2005).

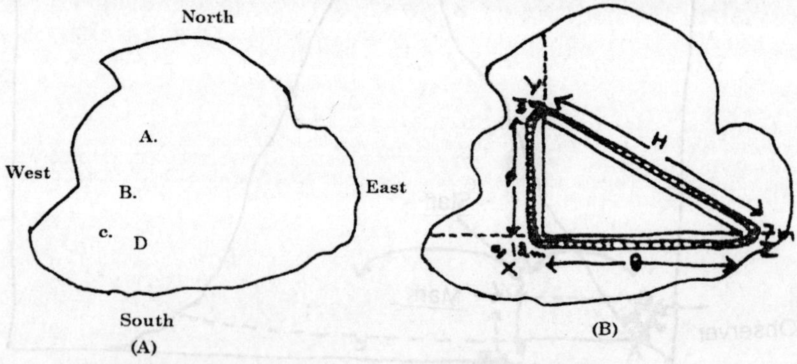


**Figure 9.** Laying out of straight line in field. (Adopted from Panwar and Bhardwaj, 2005)

## PREPARATION OF SITE

Preparing of site for planting is necessary to create congenial condition for the growth of the plants both below ground- by removing the roots and reducing the compaction of the soil and above ground – by removing the weeds and plants which may check the sunlight and reduce the growing space for the desired species.

Site preparation for planting of trees includes following steps:

i. Removing of shrubs, bushes and unwanted trees from the selected site. Tools that are generally used for cleaning of the planting sites are shown in Fig. 2a, b, e, f, g & 3 c.

ii. Marking of place of plant – After deciding the spacing between plants & pattern of planting viz square, rectangular, triangular and quincunx, marking of the point where plant is to be raised is done by pegs or by "lime".

iii. At the marked place soil working is done which depends upon terrain, soil condition and climatic conditions prevailing at the site.

## SPECIAL TECHNIQUES FOR DIFFICULT SITES

| Difficult site | Special techniques |
|---|---|
| Shifting sand dunes | Rooted cuttings of 60-80 cm length are set into the sand with the aid of an iron bar atleast 50-70 cm deep. Use of bitumen emulsion to fix sandune in initial stage of plantation is also practiced in Middle East and Ukraine. Palisading with brush wood of locally growing bushes such as *Crotalaria burnia*, *Calotropis* species buried in sand and running along and parallel to the crest of dunes also checks shifting of sand. |
| Saline and alkaline soils | Breaking of kanker by digging of pits up to 120 cms. Trench ridge (Fig. 10) method of soil working is advocated. Addition of 3.5 kg of gypsum and 2.5 kg of fym per pit helps in better survival and growth of plants. |
| Ravine lands | Gully plugging with mechanical measures, contour trenching, check dam constructions, diverting runoff |

| | |
|---|---|
| | water by making contour ridges to check flow velocity etc. should precede the planting of trees. A combination of grass and trees gives good results. |
| Cold desert | The low temperature inhibits the absorption of water by the plants roots and also causes dryness of the atmosphere. To provide moisture trench-cum-pit (Fig. 11) type of soil working techniques in sloping areas with stony surface and Irrigation-cum-drainage (Fig. 12) type of soil working is used in low lying areas (near river banks and marshy lands). |
| Lateritic soils | These soils are found in plateau areas where soil are low in fertility, have hard vesicular structure and deficient in moisture. In these areas saucer pits (Fig. 13) and rectangular ridge (Fig. 14) methods on plateau areas and contour trenches in other areas are suitable. Addition of single super phosphate and ammonium sulphate is recommended during planting of trees. |
| Dry and rocky areas | Soil almost absent, aim is to conserve moisture and soil. Contour trenches are recommended where soil is available. Areas where soil is absent pits up to 100 cm arc dug and arc filled with soil brought from elsewhere or mole hill method is adopted. In it 80 x 80 cm of area is loosened to a depth of 20 cm where seedling is planted. On uphill side trench of 80 cm wide and 25 cm deep is opened to collect water. The soil obtained from trench is heaped-up in the pit below by making a mound of 60 x 60 cm wide and 20 cm high above ground surface. |
| Wet soils | Owing to water logging, problems like lack of aeration in root zone, poor nutrient absorption and excessive salt concentration prevails. In such situations deep cross drains are prepared to drain the water from area or ridges or raised mounds are prepared. Tall planting stock gives good results. |
| Landslips | These areas are highly unstable and voluminous. Most of the time they are responsible for road blockage and sometimes destroying the roads altogether. Due to their unstable nature planting is |

difficult. To stabilize it retaining wall is constructed, however, it is costly. Alternatively, the mass of sliding area is divided into small sections at regular distance by pegging 1.5 m long green posts of species like salix, poplar. These are buried about 0.5 – 0.6 cm deep at 2-3 m apart. These posts are framed with wooden railing behind which twigs and branches are spread to hold loose soil (Fig. 15). Simultaneously in between two rows of posts quick establishing grasses are grown. The posts will also establish and become tree in few years thereby stabilizing the loose soil. The operation of stabilization should start from uphill side.

Mined areas Characters of mined land varies from bouldery to lateritic heaps, sandy to alkaline, calcarious etc, depending upon the type of mineral/ore extracted. In general they are nutrient poor, lack fertile soil and are highly unstable. In addition to adopting technique given for stabilizing landslips, addition of fertile soil, micorrhizal inoculation, addition of fertilizers and fym, robust planting stock are the common procedure which should be adopted (see Bhardwaj and Panwar, 2000).

## METHODS OF PLANTING TREES

Whether the planting to be done is in hilly area or in plain area, the method of planting are generally of three types which mainly depends upon the type of soil and kind of seedlings to be raised. Fig. 16 A-E shows the steps in planting the seedlings in areas where soil is light and with less stone. In such areas shovel is sufficient for working the soil. When the seedling stock is such that it has single tap root and can accommodate in small hole or unrooted poles are to be planted, crow bar/planting bars are used. Method of planting through planting bar is given in Fig. 17 A-F. Soils, which are hard and have stones, hoe or mattock is used. Fig. 18 A-D shows the way of planting seedling using mattock or hoe.

**Figure 10.** Trench ridge

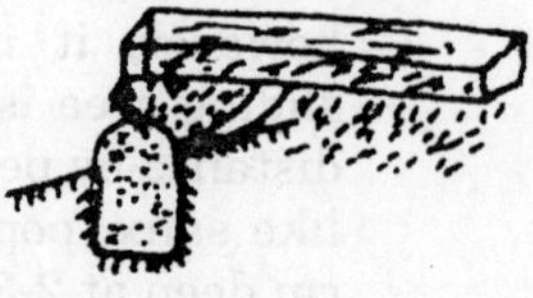

**Figure 11.** Trench-cum-pit

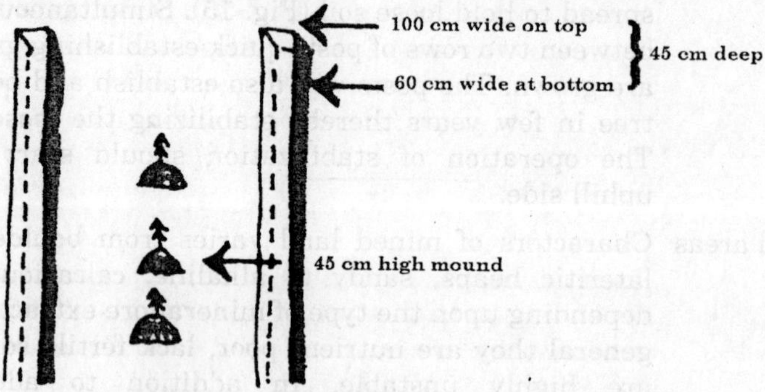


**Figure 12.** Irrigation-cum-drainage

**Figure 13.** Saucer pit

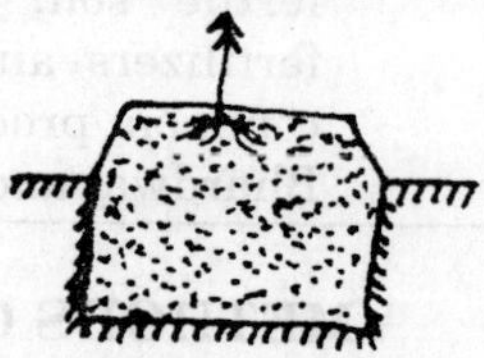

**Figure 14.** Rectangular ridge

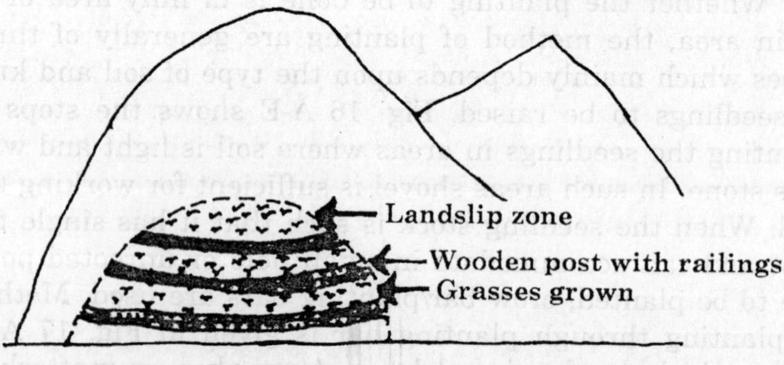


**Figure 15.** Techniques of stabilizing landslips

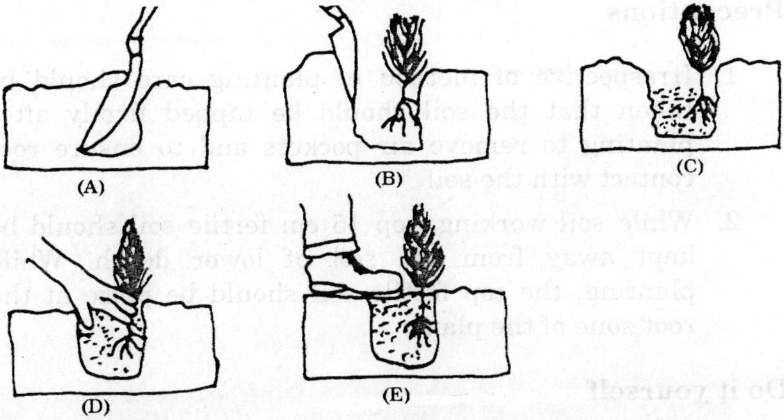


**Figure 16.** Method of planting through shovel

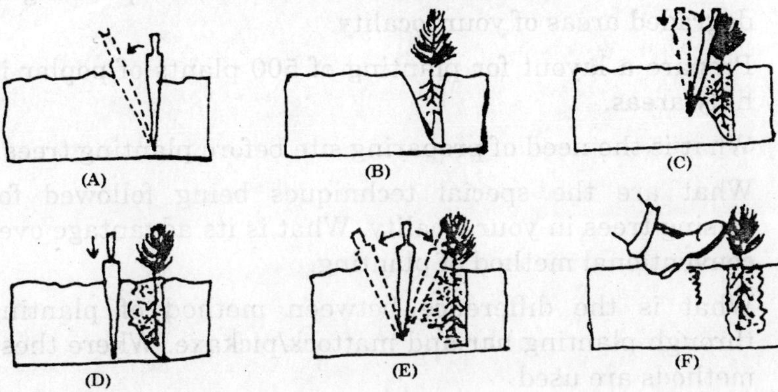


**Figure 17.** Method of planting through planting bar.

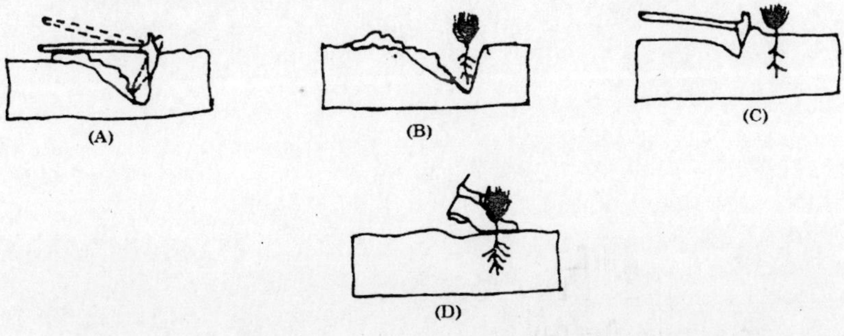


**Figure 18.** Method of planting through pickaxe or hoe.

## Precautions

1. Irrespective of method of planting care should be taken that the soil should be tapped firmly after planting to remove air pockets and to ensure root contact with the soil
2. While soil working, top 15 cm fertile soil should be kept away from the soil of lower depth. While planting, the top fertile soil should be place at the root zone of the plant.

## Do it yourself

1. Enlist 5 tree species which are suitable for planting in degraded areas of your locality.
2. Prepare a layout for planting of 500 plants of poplar in hilly areas.
3. What is the need of preparing site before planting trees.
4. What are the special techniques being followed for raising trees in your locality. What is its advantage over conventional method of planting.
5. What is the difference between method of planting through planting bar and mattock/pickaxe. Where these methods are used.

Chapter **4**

# Tending Operations

The aim of a forester is to obtain good quality and maximum quantity of wood per unit area. To achieve this goal efforts can be diverted either toward site improvement and/or the crop itself. The treatments that are done on the crop are known as tending operations. It is defined as "An operation carried out for the benefit of a forest crop, at any stage of its life between seedling and the mature stage: it essentially covers operations both on the crop itself and on the competing vegetation e.g. weeding, cleaning, thinning, pruning, climber cutting etc, but does not include regeneration felling and ground operations like soil working, drainage, irrigation and controlled burning." Tending includes weeding, cleaning, pruning, thinning, climber cutting etc. Among these most important operations are pruning and thinning.

## Pruning the trees

When the plant is of 3-4 years, it starts putting large branches to increase its surface area and hence its photosynthate production. However, the base of these branches form knots in timber. If the wood in the trunk grows around the branches, when branches are living, the knots formed are tight and will not necessarily weaken the timber. If the branches die on the tree, the new wood grows around the dead branches. When the sleeper is sawn from the tree, these knots formed from the dead branches are called loose knots or bark encased knots and are likely to fall out and leave a hole in the timber. For obtaining high quality timber it is desired that it should be free from knots. Thus removal of all branches from initial stage allows a forester to get timber of desired quality. The condition of pruned and unpruned log is shown in Fig. 19 A-C. High quality wood have their end use for furniture, veneer or plywood. In addition to improving the quality and value of the wood. Pruning also:

- Make hygienic condition in the forest
- Helps easy movement of machineries in the forest
- Helps in controlling fire
- In agroforestry, allows more light to reach the ground for the growth of associated crops.

## PRUNING PRINCIPLES AND TECHNIQUES

1. Prune when the trees are small to avoid wastage of timber. Fig. 19 A-C shows cross section of variation in amount of quality timber obtained from a tree, which had been pruned at different ages. Log shown at Fig. 19C was pruned from initial stages hence more amount of quality wood can be recovered. Log at Fig. 19B was pruned after 4 years of age hence the quality of log shown within dotted lines was poor. The tree that was not pruned throughout its life is shown in Fig. 19A. It will give the poorest quality of wood due to presence of knots.

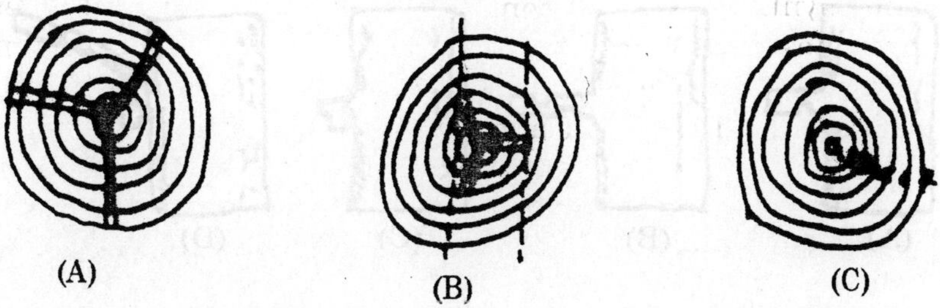


**Figure 19.** Effect of pruning stage on amount of timber obtained

2. Use saw instead of axe or sickle. Fig. 20 a-b shows the variation in quality of timber obtained, in latter years, after using axe and saw.

**(A) Pruned by axe** **(B) Pruned by saw**

**Figure 20.** Variation in wood quality by use of axe and saw in pruning

3. Winter season is the suitable time to prune the branches
4. In good pruning operations, the object of the operation is to remove the whorls of branches as close to the tree stem as possible without leaving protruding branch stems or 'coat hangers'. 'Coat hangers' result in encased knots and bark pockets which reduce timber value.
5. While pruning, to avoid stripping of bark, a under cut should be given. When the branch is very large and heavy first branch should be cut ½ feet away from the trunk of the tree by using undercut and then close to the trunk (Fig. 21 a –d)

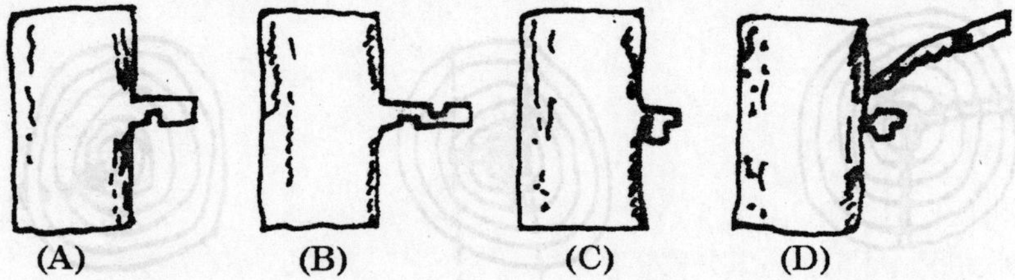


**Figure 21.** Method of pruning heavy branches

6. For hardwoods, the branch should be cut with a slight angle (Fig. 22 a-b).

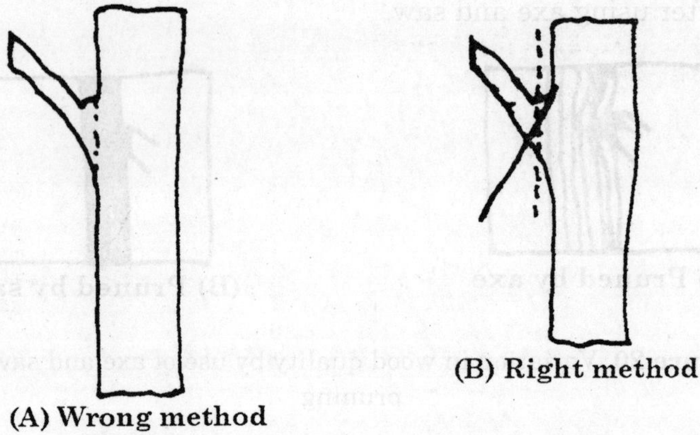


**Figure 22.** Method of pruning hard wood.

7. Trees tend to grow straight up if leader is free from competition. To achieve this, competing branches need to be removed. On non competing branches can be kept to side opposite to any lean help leader grow straight (Fig. 23 a-b).
8. Remove heavy side branches or the branches, which had grown too long (Fig. 24). Remove leaders that have become deformed or have been killed by frost (Fig. 25).

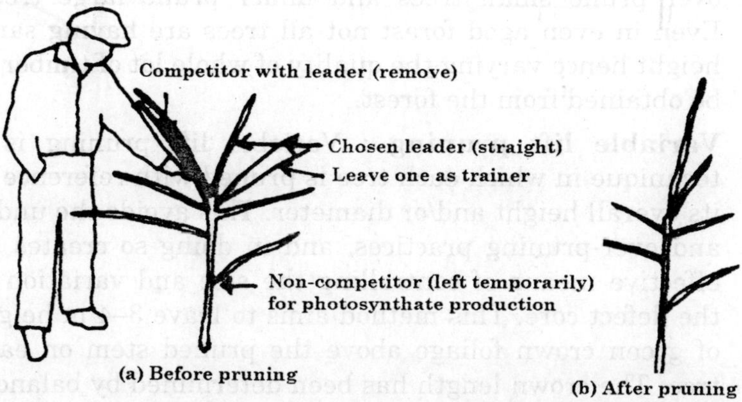


**Figure 23.** Plant before and after pruning

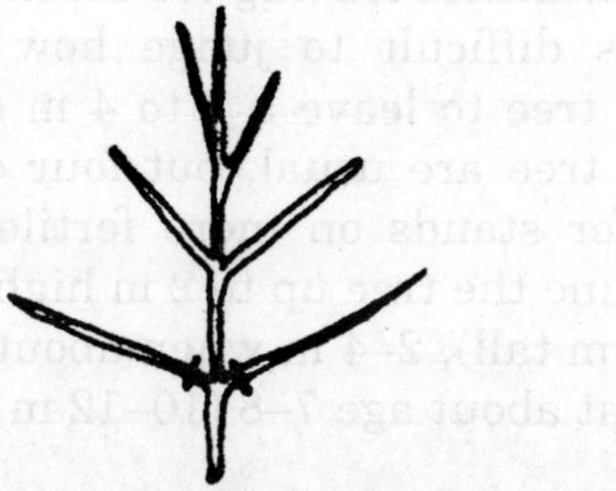

**Figure 24.** Prune extra long side branches

**Figure 25.** Prune deformed leader

## Methods of pruning

There are two ways of pruning a whole forest viz. Fixed height pruning and variable lift pruning.

a. **Fixed height pruning** : In this method, for whole of the forest a specific height is fixed say 2 or 3 m depending upon age of the average crop. All branches which occurs in this fixed range are removed irrespective to the height of individual trees. This method is suitable for even aged forest, however, in selection forest using this method will over prune small trees and under prune large trees. Even in even aged forest not all trees are having same height hence varying the quality of whole lot of timber to be obtained from the forest.

b. **Variable lift pruning** : Variable lift pruning is a technique in which each tree is pruned with reference to its overall height and/or diameter. This avoids the under and over-pruning practices, and in doing so creates an effective means of controlling the size and variation of the defect core. This method aims to leave 3–4 m height of green crown foliage above the pruned stem on each tree. The crown length has been determined by balances between the removal of too much crown (which slows the tree growth) and the removal of too little crown (which promotes the growth of branches leaving too much defect core). In practice, it is difficult to judge how much pruning is needed for a tree to leave a 3 to 4 m crown. Three pruning lifts per tree are usual, but four or five lifts may be required for stands on more fertile land. Typical practice is to prune the tree up to 2 m high when it is about age 4–5 (5–6 m tall), 2–4 m when about age 6 (7–8 m tall), and 4–6 m at about age 7–8 (10–12 m tall).

## THINNING OF PLANTS

It is wasteful and uneconomic to prune defective trees. If possible, any defective trees should be removed as soon as they are identified, so leaving more growing space for the better trees. Practice of removing of these defective trees is known as thinning. It is defined as "Felling made in an immature stand for the purpose of improving the growth and form of the trees that remain, without permanently breaking the canopy". Thinning is the removal of some trees from a stand of trees. The removed trees are of poor quality. This means that the remaining trees will have less competition for available resources and the best trees will be nurtured to their full potential. Thinning enables foresters to:

- Utilize the volume which would otherwise have lost during competition.
- Select larger, straighter and healthy trees
- Select trees with small branching habit
- Improve/maintain stand composition and regeneration condition of forest.
- Increase tree size, and decrease the growth time to reach a marketable size.
- Increase financial out-turn by producing large sized timber.
- Allow for the potential growth of the best looking trees by producing more space between them.
- Improve hygienic condition in the forest.

**When to thin the crop**

Factors that influence optimum harvest time are species, growth rate and type of market. These will determine if the plantation is to be thinned during its rotation. Eucalyptus for example, is mostly grown for pulpwood and is normally clear felled at an early age without any prior thinning. Teak on the other hand is usually thinned several times before the final harvest. Another factor may be slope. For example on steep slopes thinning of species may not be economical due to high cost of harvesting and production of less volumes as thinned material. In such areas trees may be left unthinned until final harvest when larger volumes will be obtained. Leaving trees unthinned for a whole rotation of 20 or more years is generally an easy but hazardous choice. Unthinned trees compete as they grow which stresses the trees and restricts their diameter growth. This makes them more vulnerable to disease and death. Light Non-commercial thinning may be an option in these circumstances. In extremely windy areas, plantations may be prone to toppling after thinning. In such areas, thinning should be light and carried out before the plantation becomes too tall. However, as a general rule the first thinning is best carried out during the first pruning stage. The second and subsequent thinnings may be carried out at the same time as medium and high pruning. The final thinning can be delayed if:

- The trees to be removed are required to control branch sizes;
- The trees to be removed are to be used for posts, poles, pulpwood, etc
- Extreme weed problems are present.

**How much to thin**

The ground area available to each tree influences its growth. Thinning regimes are calculated on a trees-per-hectare basis. Thinning may not result in even distances between trees, however trees should not be too close (2.5 m), otherwise competition will restrict growth. On the other hand, spacing between trees greater than 12 m apart are likely to underuse the available area and waste valuable land.

**Selection of tree for thinning**

**(A) On the basis of the quality of log**

The quality of a tree is assessed by the characteristics of the butt log (the log cut from the lowest part of the tree trunk) and the second log (immediately above the butt log). The butt log provides the highest financial return from the tree, especially if it is pruned. The bottom two logs contain practically all of the tree's commercial value. To determine which of the trees in a forest will be thinned out, the potential quality of the timber that a tree will ultimately produce must be assessed. The quality of the butt log, can be confidently assessed when the tree is about 6 m tall. The quality of the second log, the log cut directly above the butt log, can be assessed when the tree is approximately 12 m tall. Thus the trees, which will produce more quantity and quality of these two logs, are retained and rest are removed.

**(B) On the basis of crown spread**

Since the widest part of the tree is its crown and for better growth of plant sufficient space has to provided for better crown development. Thus crown can be used as a mean of thinning the forest crops. On the basis of crown freedom, trees are classified as :

Spaced : Trees which have spread of free crown equal to or more than ¾ of average spacing between trees before thinning.

Co-spaced: Trees which have a spread of free crown < ¾ but more then ½ of average spacing between trees before thinning.

Sub-spaced: Trees which have a spread of free crown between ¼ to ½ of average spacing between trees before thinning.

Mal-spaced: Trees which have a spread of free crown <1/4 of spacing between trees before thinning.

De-spaced: Trees which have no free crown.

Average spacing/ crown spread is calculated as square root of average space occupied by tree i.e

$$\text{Crown spread} = \sqrt{43{,}660/N} \text{ in feet}$$

Or

$$= 100/\sqrt{N} \text{ in meters}$$

(where N is number of trees/acre or hectare)

**Example :** Let there are 9000 trees growing in a acre. Find out crown spread between the trees.

$$\text{Crown spread} = \sqrt{43{,}660/N} \text{ in feet}$$

$$= \sqrt{43{,}660/9000}$$

$$= \sqrt{4.851}$$

$$= 2.20 \text{ feet}$$

**Example:** Taking 2.20 feet as crown spread, calculate which trees will be spaced, co-spaced, sub-spaced and mal-spaced.

Spaced trees: Where free crown is between ¾ to average spacing/crown spread.: In present case average crown spread is 2.20 feet. Its ¾ value will be ¾ x 2.20 = 1.65 feet. Thus trees having crown spread between 1.65 feet to 2.20 feet are spaced trees.

Co-spaced trees: Crown spread between ½ - ¾ of average crown spread: For present case ½ x 2.20 = 1.10 feet and ¾ x 2.20 = 1.65 feet i.e tree having crown spread between 1.10 to 1.65 feet are co-spaced.

Sub-spaced: Crown spread between ¼ - ½ of average crown spread: For present case ¼ x 2.20 and ½ x 2.20 i.e trees having crown spread between 0.55 to 1.10 feet are subspaced.

Mal-spaced: Crown spread less then ¼ of average crown spread: For present case ¼ x 2.20 i.e the trees having crown spread <0.55 feet are mal-spaced trees.

Thus on the basis of this classification trees are thinned. Malspaced and subspaced trees are thinned to provide sufficient growing space for rest of the trees.

**(C) On the basis of the height of the trees**

**On the basis of height, trees are classified as**

1. Dominant trees (D) : Trees which form upper most canopy of forest. It has two sub division:

Predominant trees ($D_1$): Tallest trees which are free from vertical competition.

Codominant trees ($D_2$) : Rest of the dominants which are about 5/6th of the average height of predominants.

2. Dominated trees (d) : Do not form upper most leaf canopy but their leading shoots are not topped by neighboring trees. Their height is about ¾ of tallest trees.
3. Suppressed trees (s) : Trees which reach only ½ to 5/8th of the height of tallest trees. Their leading shoots are overtopped by neighbors.
4. Dead and Moribund trees (m) : Trees which are bent over and badly leaning.
5. Diseased trees (k) : Trees infected with parasites and are seriously affected and also pose danger to its neighbor.

Consider Fig. 26, let it be the situation in a forest (adopted from Panwar and Bhardwaj, 2005). In such forest depending upon the tree classification light thinning advocates removal of diseased (k), dead and moribund trees (m). If thinning to be done is moderate then in addition to k and m category of trees, suppressed (s) trees are also removed. In heavy thinning some of the codominant (CD) and dominants are also removed. However, the removal of these trees depends altogether on the experience of the forester. Thus it is called as subjective thinning.

Crown thinning also called as French thinning or high thinning and ordinary thinning also called as German thinning or low thinning are examples of subjective thinning. However, it should be noted that the crown thinning is applied in species which are moderately shade tolerant, whereas, ordinary thinning is applied in light demander species. Thus while applying crown thinning on deodar (*Cedrus deodara*), trees which are predominant and codominant and dead and diseased are removed (Fig. 26).

Conversely on chir pine (*Pinus roxburghii*) or Gamar (*Gmelina arborea*) ordinary thinning will be applied and besides dead and diseased trees suppressed (s) trees will be removed (Fig. 26).

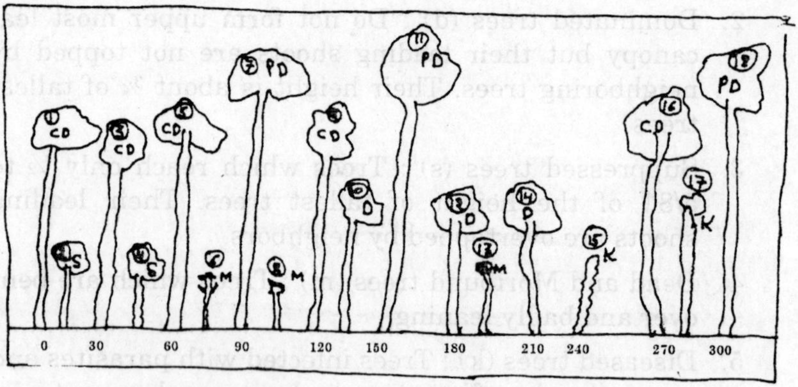


**Figure 26.** Natural forest depicting trees of different classes (Adopted from Panwar and Bhardwaj, 2005)

## HARVESTING OF CROPS

Harvesting is the culmination of plantation and it determines the results of the management which was done in past. The amount of return that will be obtained from the plantation depends largely upon the harvesting procedures. If volume growing in the forest is sufficient and of good quality, but while harvesting and conversion the logs gets cracks/splits, the market value of that log reduces and all efforts put in past goes waste. To avoid these exigencies proper harvest planning procedures are adopted. These includes:

a. Preparation of topographic maps showing location of compartments within forest.

b. Record of contour maps to plan location of roads for extraction of harvested material and moving of machineries.

c. Arranging trained labour for harvesting trees.

d. Decision regarding manual or mechanized harvesting.

## Techniques of harvesting

After the rotation age is over trees are felled. Trees are either felled manually or by mechanical means. To avoid split butts, snapped stems and fractures and to give direction to the felling of tree, some techniques are adopted which are as follows:

### Felling using axe

Fig. 27 A shows the method of felling the trees using axe. If the tree is to be felled in east direction, first an under cut is given on the eastern side of the tree. The depth of under cut should vary with the size of the tree and whether or not it is to be felled in the direction of its lean. In straight even-crowned, and sound trees the bottom of the undercut is chopped or sawn horizontally for a distance of 1/4$^{th}$ to 1/3$^{rd}$ of the diameter of the tree. When the tree is leaning in the direction of fall, the undercut should be deeper to prevent it from splitting. Corner notches should also be chopped in on each side of the undercut in leaning trees to prevent pulling side splinters as the tree falls. On exactly opposite side of the under cut the upper cut is made, 2-3 inches above the under cut, to fell the tree.

### Felling using axe and saw

Fig. 27 B shows the method of felling tree using axe and saw. Through axe the under cut is made and from exactly opposite side saw is used for dislodging the tree from its base. The wood loss in this method is lesser as compared to use of axe alone

### Felling using saw

Fig. 27C shows the method of felling tree using saw alone. While using saw alone a wedge is required to protect the saw from being jammed by the weight of the tree. The wood loss in this method is least compared to above two methods. These days' power chain saws are also available which are speedy and waste of wood is also less (Fig. 3d).

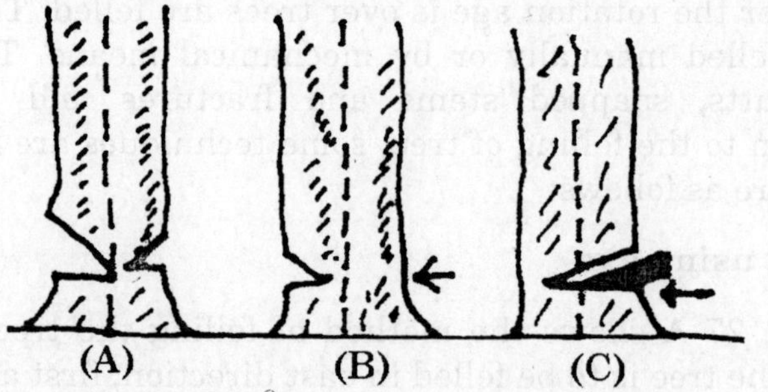


**Figure 27.** Methods of harvesting trees

**Technique of making under cut**

The aim of making an under cut is to preventing the splitting of the log as well as to give the direction of felling of the tree. If the under cut is not made proper wastage of wood is more and tree may become useless. The accurate shape and depth of under cut is shown in Fig. 28 A. If the under cut is made broader (Fig. 28 B) it will waste the valuable wood of the butt log. If the under cut is made narrower (Fig. 28 C) or if the slanting part is not joined with the horizontal part of the undercut (Fig. 28 D) the tree will not fall in the proper direction.

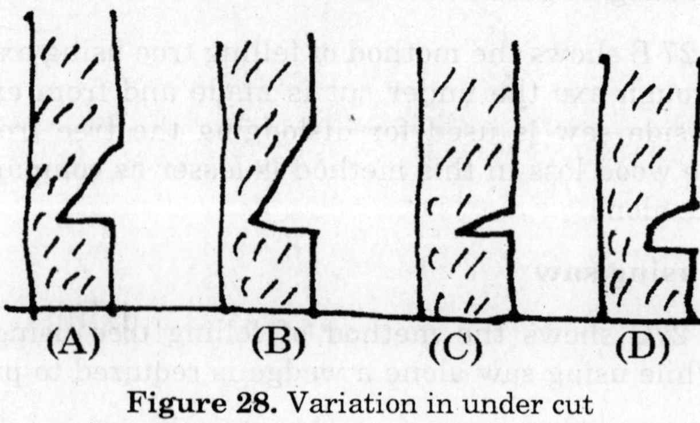


**Figure 28.** Variation in under cut

**Method of felling the trees in different localities**

The best place for felling a tree is a leveled area with no stones. However such areas are seldom available at the harvest sites. However, wherever, possible while harvesting the trees big stones of logs lying near the tree, which is to be harvested, should be removed. Consider Fig. 29A, the log/ stone lying on the forest floor will result in breaking of the tree due to variation of weight on the both side of the stone/log. Consider Fig. 29 B such situations are acquainted in undulating lands. Felling in such situation will break the tree. Hence the tree should be felled in reverse direction. In hills felling of tree toward down slope will damage the tree, however, if the tree is felled upslope then it will be difficult to remove it from its position due to entangling of the stump of the tree and the base of the dislodged tree due to weight of the tree. Hence in hilly areas the tree should be felled at 45° angle of the slope (Fig. 29C).

(A)
(B)
(C)

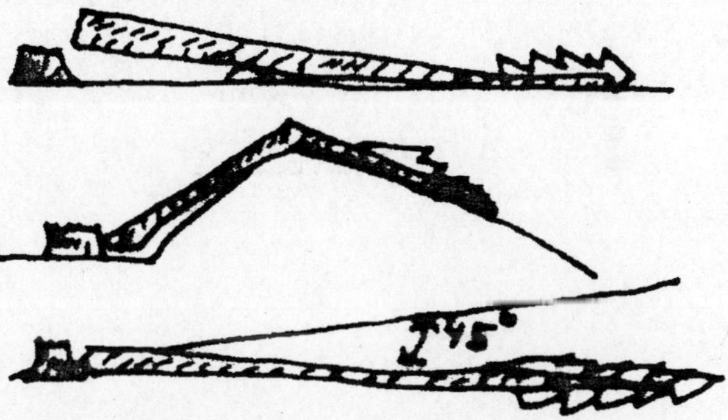


**Figure 29.** Methods of felling trees

**Do it yourself**

1. What is the need of pruning the trees?
2. What is the accurate stage of pruning the trees.
3. Enlist the tools used for pruning.

4. What is the advantage of variable lift pruning over fixed pruning.
5. What is the difference between thinning and pruning.
6. Which trees are dominated and suppressed trees, with reference to height, if height of predominant tree is 20 m.
7. Find out the trees which are cospaced and malspaced in a forest if the number of trees per hactare are 700.
8. What is the stage of crop when it should be thinned.

# Chapter 5

# Plantation Planning

To be successful in plantation enterprise, exercise on what to plant, where to plant, how much to plant etc should start prior to the acquiring of the land for the said purpose. The decision made in advance will reduce the burden and exigencies which otherwise are possible. This will also help the forester and/or owner to keep himself prepared for obvious problems which are endemic to plantations like insect-pest attack, fire hazards etc. In the following section the points which are to kept in mind while going for plantation are given in a ordeal manner. However, some additional points can also be considered depending upon the locality and aim of the owner.

## Steps while planning the plantation

### DECIDE WHY TO START A PLANTATION

This decision is a very important decision. The answer to this decision will open the way for different kinds of requirements and methods to establish plantation. There can be number of answers to this question e.g. for giving aesthetic look to the area for tourism purpose or to improve fertility status of the soil or to obtain profitable timber or conserve wildlife or biodiversity or to reduce pollution etc. To find the answer of this, three questions need to be answered viz:

**(a) What are long term goals**

These long-term goals will decide what you want most from your land. Long-term goals may be timber harvest, diversity conservation, aesthetic, reducing soil erosion. More these goals are specific more useful they will be in decision-making.

**(b) What are short term goals**

While the plantation is growing what do you desire from the land? Whether firewood or agriculture crops or picnic spot etc. Answer to this question will refine the decisions with regards to designing and maintenance of the plantation. E.g. if short-term goal is getting agriculture crops so initial spacing between trees would be larger. If goal is firewood production then initial spacing would be close.

**(c) How much time and money can be spared**

This will lead to a realistic decision-making. If time in hand is less, short rotation crops are to be selected. If time is less mechanization is a good option. If both time and money is less to suppress weeds or prune branches then such species should be chosen which are good competitor and have self-pruning ability.

## CHOOSE THE APPROPRIATE SPECIES

After deciding what you want from your land the next most basic question needs to be answered is which species will meet the forestry goals and which species is suitable for the locality (edaphic and climatic).

**(a) Which species will meet forestry goals**

On the basis of owner and market demands and value of the species (with reference to timber, fuel, fodder, soil conservation) few species need to be considered first and they are ranked on the basis of their priority.

**(b) Match species and site**

On the basis of the character of the the site viz soil (texture, structure, fertility, pH, depth, aspect, slope, moisture); climate (precipitation, frost, snow, temperature, altitude); disease and insect incidences (past history of disease prevalence or incidences in nearby locality) and animal damage (grazing, trampling, borrowing, rodents etc), the species chosen at 2a above are screened for their suitability at the choosen site. It should be borne in mind that some site characters like fertility, structure, insect-pest and animal damage can be controlled by the foresters, however, slope, aspects, precipitation, texture etc cannot be changed. Thus the factors which can be controlled may be negated while choosing the species for plantation.

## PREPARATION OF SITE

Once the species is decided, the site preparation is done according to the suitability of the species. It should be borne in mind that good seedling establishment depends on good site preparation. Good site preparation will reduce competition, prepare proper soil bed and improve drainage. There can be number of approaches for preparing site like mechanical/manual means of soil working, chemical or manual means of eradicating weeds. The choice depends upon availability of funds and time in hand. However, some endemic problems do exist which don't allow free choice. Eg.

in hilly areas soil working is not possible by mechanized way hence dependence is altogether upon manual labour or in areas where human habitation is near, chemical spraying may be hazardous to the people. In these areas weeds are to be removed by other means available with forester/ owner, which may be costly and time consuming. While preparing site as far as possible, soil should be exposed to the least otherwise there would be heavy loss of fertile soil and invasion of weeds.

## CHOICE OF PLANTING STOCK

Depending upon the prevailing site condition and extent to which site has been prepared, the planting stock is chosen. Planting stock is chosen on the basis of its age and dimension. While choosing planting stock on the basis of age, stock may be of 2+0 or 2+1 etc. where 2 represents the number of years the seedling was kept in nursery beds and 0 or one denotes the number of years the seedling was in transplant beds. Thus 2+1 seedling is of 3 years and was kept in nursery beds for 2 years and in transplant beds for 1 year. In general younger seedling requires best site, whereas, older seedlings can adapt in poorer site. Also the seedlings, which grow in transplant beds, will have more fibrous root and hence good adaptation. However, it should be kept in mind that it is expensive to produce and plant large sized seedlings or older seedlings. Choice also need to be made between containerized or bare root seedling, which depends upon the species chosen for plantation as some species don't respond when grown bare root. However, containerized seedlings are more expensive then bare root seedlings.

## DESIGNING THE PLANTATION LAYOUT

Layout of the plantation is the last major step while planning the plantation. In this all the decision made earlier with reference to choice of species and site, site preparation and choice of planting stock are clubbed and matched with the objectives, to decide – access roads, fire breaks, erosion control, fencing needs, harvest plan, number of years, trees to be ordered and planted, their arrangement etc. The layout should be prepared on the paper and depending upon the locality factors the problem, which can occur at present layout, can be rectified. It should be kept in mind that layout at this stage can be easily modified, a plantation once set cannot be changed. This layout also helps in estimating the total cost which is required to be invested, that too at different intervals. On the basis of the objective of owner the layout of the plantation will differ. Following are some of the example depicting change in layout with change in owner's objective:

**(i) If objective is Timber production**

**Considerations while producing timber**

— Choose species and clones which have timber value and uniform growth and have ready local market. Monocropping is best, exotics can be introduced.
— Space the trees in a systematic manner with well defined access roads and fire breaks. Also leave space for power lines etc.
— Go for extensive pruning to improve quality of wood.
— Weeding and thinning is utmost important to reduce competition.
— Avoid steep slopes, wet lands, they will not give uniform growth.

Taking these factors into account the layout for timber production is shown in Fig. 30.

* For getting timber production year after year (sustained yield), the land has to be divided in such a way that each year an equal area is harvested. This is possible

by planting equal area each year up to the rotation age of the plant (For detail see Khanna—Forest management).

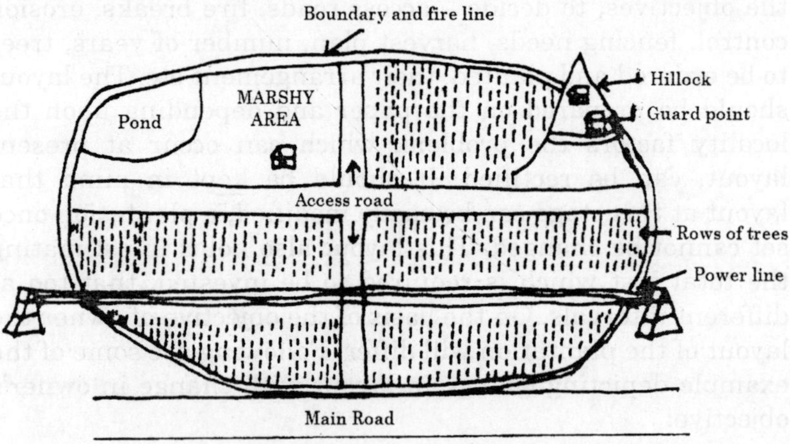


**Figure 30.** Layout for the purpose of timber production

### (ii) If objective is Wild life management and Tourism

### Considerations for wildlife management and tourism

- — Determine the kind of local fauna present and according to their breeding and feeding habit provide open and dense space.
- — Increase biodiversity of plant to fulfill the need of different herbivores. The population and variety of herbivores determines the survival of carnivores and hence sustainability of the forest. Prefer indigenous species of both flora and fauna.
- — Create irregular boundaries to provide more edge for game and tourism.
- — Fire breaks and boundary closure to prevent domestic animals to enter inside.

While considering these points the layout of the plantation for the said purpose is shown in Fig. 31.

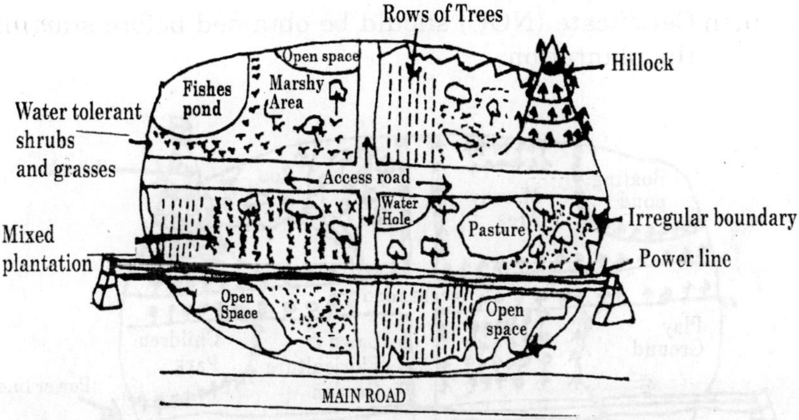


**Figure 31.** Layout for the purpose of wildlife management and tourism.

### (iii) Aesthetic and/or picnic spot

### Considerations for aesthetic and picnic spot

— Vary layout and location of planting-like dense plantation, open space, rows along road.

— Choose species, which are flowering and have variety of crown shapes like oval, conical, round, irregular etc.

— Mix short statured and tall trees.

— Provide facilities like boating, plant water tolerant trees near pond.

— Keep some wildlife in cages and old trees are retained, as they are shelter of number of birds.

The layout for the plantation for aesthetic and/or picnic purpose is shown in Fig. 32.

⇒ In the above examples it was shown that the same land could be utilized for different plantation arrangement based upon owners objective. Once layout had been finalized, this has to be passed through the local authority and a No Objection

Certificate (NOC) should be obtained before starting the plantation.

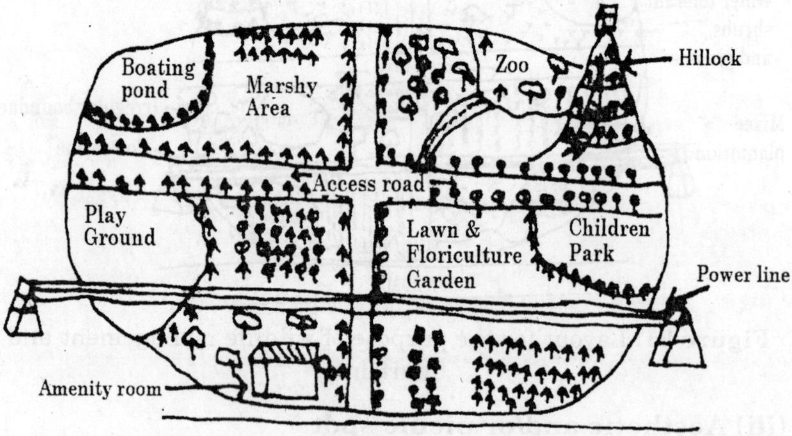


**Figure 32.** Layout for the purpose of aesthetic/recreation.

**Do it yourself**

1. Prepare a harvest plan for 1 hac. area of which ½ hactare area is hilly and rest is leveled/plain.
2. What method you will use to harvest a tree in hilly area against the lean of the tree.
3. Calculate the percent loss of wood while using saw alone, saw + axe and axe alone.

Chapter **6**

# Impact of Plantation Forestry

Planting trees is an age-old practice. However, after 1960's trend of launching large scale plantation picked up tremendously, as international donor supported such plantation for producing raw material for paper, pulp and timber. It is estimated that such large scale monoculture are now increasing at a rate of 4.5 million hactare per year of which 79 per cent is planted in Asia. India significantly contributes to these plantation and ranks among first five nations of world. Plantation forestry though a recent phenomenon, but has been a very debatable issue. In following sections some of these points has been discussed.

## PLANTATION FORESTRY AND BIODIVERSITY

There are two school of thoughts with regards to effect of plantations on biodiversity. If a large area of natural forest is felled to raise plantation, there will be a loss of biodiversity. The same applies when a natural savanna ecosystem is replaced by a plantation of alien species, as has frequently happened in South Africa, Uruguay and

Argentina. It had also been observed that larger the area of natural forest converted to new plantation, the greater is the number of species which are affected. This means that large scale monoculture of plantations are detrimental to the biodiversity of the region. Advocators of plantation forestry are of the view that it is the location of the plantation, its size, length of rotation etc which depicts the plantation affect on biodiversity, and not the plantation concept as a whole. e.g. A plantation raised after replacing natural forest or savanna will definitely lead to loss of biodiversity but if similar plantation is established on degraded land, might bring an increase in biodiversity. As a general rule, biodiversity is at its richest in the humid tropics, or close to sea level, and declines on a continuum towards the poles and from sea level to high altitude. If a plantation replaces primary forest in, say, Costa Rica or Indonesia, it will lead to a greater loss of biodiversity than it would in southern Chile or Canada. Moreover, as per the global trend the conversion of natural forest to raise plantation accounts for only 6-7 per cent of total forest lost, whereas, the rest 93 – 94 per cent loss of the natural forest is due to increasing encroachment for agriculture and industrial development. With regards to India and China, it may be stated that at present plantations are increasing the biodiversity as these are raised only on the degraded and abandoned agricultural lands. However, to bridge the gap between demand and supply of wood and simultaneously protect biodiversity, retaining a natural corridor between blocks of natural forest and plantation is imporative. Through strict rules, the plantation owner can be compelled to keep a patch of land of their plantation maintained for biodiversity conservation. Using wood that comes from plantation and not from the natural forest can also protect Biodiversity as is in New Zealand where 99 per cent of wood consumed comes from plantations. Such rules will reduce pressure on the natural forest.

## PLANTATION FORESTRY AND SOIL

It has been observed that plantation raised on short rotation basis are nutrient exhaustive and make the land more prone to soil erosion, particularly during the initial stages when they are young. Also since their management is such that very little under growth is allowed hence soil erosion is more. Moreover use of heavy machineries during felling and extraction also compact the soil. These problems are because of poor management practices in plantations otherwise plantations don't have that adverse effect on soil. If site preparation techniques like contour planting, construction of micro-catchments are adopted erosion of soil can be checked. While comparing plantations with agriculture crops it had been found that plantations are generally much less degrading to the soil than many commercial agricultural crops. A study conducted by FAO, 2000 found that the amount of nitrogen removed by a cereal crop was two and half times more than the amount removed by a eucalypt plantation. In the case of phosphorus it was fifteen times more. A study comparing fast-growing pine with grains and potatoes found that the latter removed 15 to 35 times more nitrogen, 80 to 250 times more phosphorus, and up to 10 times more calcium than the fast growing pines (Will and Ballard, 1976). Fast growing plantations require much less fertiliser per hectare than many agricultural crops, and a single application per rotation will generally suffice. Need less to mention that while plantation are growing they are contributing some organic matter to soil through leaf fall however, in case of agriculture nothing is returned to soil. Thus it goes without saying that good management will help to minimise soil loss and declines in fertility, regardless of whether the land is devoted to fast-wood plantations or annual agricultural crops.

## PLANTATION FORESTRY AND SOCIETY

Establishing large-scale blocks of fast growing trees has an impact on not only environment and soil but also on the people residing nearby. It has been said that plantation generate employment only during planting and harvest and the employment generated is of unskilled labour, seasonal and with less wages. However, establishment this is the only one part of the plantation forestry, the employment generated in industry dependent on raw material obtained from plantation forestry are also to be accounted for during such calculations. The extent to which plantation can generate employment also depends upon location of plantation. If plantations are established on fertile agricultural land, the chances are that the number of jobs created will be considerably less than the number previously supported by farming, especially in developing countries. However, if plantations are established on land that has either been abandoned or has previously been of little use to farmers or others, then they may bring new jobs in that area. As a general rule, plantations grown for higher value sawn logs have a much better chance of generating local jobs than plantations grown for low-value pulp. While establishing plantations, the local peoples should also be taken into confidence to avoid any conflict and hence failure of the plantation. This can be done by providing most of the jobs, wherever possible, to the local inhabitants. Locals should be preferred while auctioning the thinned or pruned materials from the plantation. Providing facilities like school, drinking water, roads etc will also help in avoiding conflicts. Thus in may be said that when it comes to plantations forestry and employment, what holds true for one area, and one type of plantation, may not be same for another. Some plantations generate employment; others take jobs away. Also a rapport building measures between locals and plantation manager will help in avoiding conflicts. In India company/smallholder relationships have rapidly evolved, with wood being traded in the open market, and companies competing to meet their wood requirements. Outgrower schemes have benefited from a government policy that limits the area of land that can be held by a private owner. This has forced processing companies to buy most of their supplies from smallholders, and small-scale tree growing is increasingly seen as a viable land use. Local involvement may mean there is less likelihood of conflict, a common problem for many plantation owners. Plantations provide employment and income for some, and profits as well as a way to spread risks between agricultural and timber crops, for others. Companies that purchase the entire output of wood from an outgrowers' scheme sometimes take advantage of their monopoly by driving down prices. Inevitably, farmers resent this. In places where several companies are competing for the same wood supply, some farmers may get credit and technical assistance from one, and then break their agreement to sell to another. Outgrower schemes have considerable potential, and can help to address some of the environmental and social problems that currently beset large-scale projects.

## PLANTATION FORESTRY AND ECONOMICS

Large scale plantations are carried out by the private owners only because plantations are a profitable business otherwise they would not have entered in this enterprise. However, plantations may also result in economic losses. Most industrial plantations have been established with financial support from the state. This often means that public taxes are used to subsidies privately owned ventures. If subsidised plantations fail to yield tangible public benefits, then taxpayers may feel they have been short-changed. The economic costs of development may also be borne locally, rather than nationally. For example, villagers may be deprived of agricultural land, and thus experience a decline in income, or a plantation might lead to hydrological changes which adversely affect downstream crop yields. But still plantations are to be carried out because demand of wood is increasing. At present, the world population is about 6.2 billion. According to the International Programs Center of the US Bureau of the Census, it will reach 9.4 billion by the year 2050. Even if per capita consumption were to remain stable, demand for wood-based products would rise significantly. Since 1913, world paper consumption has increased 17-fold, while the population has increased by a factor of 4. According to FAO, world paper consumption will reach 443 million tonnes by 2010, an 80 per cent increase since 1990. Consumption trends therefore favour a rapid increase in demand for wood produced by fast-wood plantations (FAO, 2000).

In general forestry enterprise has picked up by the help of subsidies given by respective governments both in developed and developing countries. The amount of subsidies for plantations are more in developed countries. In countries of European Union, the average subsidy/incentives given for plantation are at the rate of $ 1,421 per hactare with an additional $ 761 per hactare for maintenance, which is many folds then those given in developing countries like India. Governments supports the decision of subsidies by saying that these will encourage more plantation which will lead to more carbon storage, soil protection, benefit to society etc. Subsidies are also justified by saying that if plantations are successful the amount given as subsidy will be recovered in terms of taxes on the produce from plantation. It is also argued that environment benefits from the plantations cannot be quantified in financial terms hence subsidies given for plantation are a good decision. However, There are many compelling arguments against the use of subsidies. They create distortions within the economy, both internationally and locally. Companies that do not receive subsidies find themselves at a competitive disadvantage to those companies which receives these subsidies. Looking into account the economic principles plantation enterprise sector is most likely to prosper in those countries where there is political and macro-economic stability; where there are open markets; where property rights are clearly established; and where the government has the ability to enforce the law (FAO, 2000).

From the ongoing discussion it may be concluded that though plantations are necessary for economic development, for meeting day-to-day demand of rural and urban population, for environment etc, however, it also have some darker sides. To reduce the negative points of plantation forestry what is required is proper planning, choice of site, species, rapport between plantation manager and local people, good economic policies etc. Now-a-days, many plantation projects under take environmental and social impact audits at an early stage of developing of plantations. We need to develop a system formula into which we could feed a mass of data when evaluating a plantation project: projections of jobs to be created; impact on wildlife; impact on hydrology and adjacent farmland; contribution to the national exchequer; benefits in terms of carbon sequestration and so forth. This would help us to come up with an answer that would tell us whether or not the plantation project should go ahead.

## QUESTIONNAIRE FOR ASSESSING IMPACT OF PLANTATION OF YOUR LOCALITY

a. Location : State ______, District________, Village________

b. Ownership of plantation : Corporate/ Government/ Individual

c. Name of the owner:________________________

d. Extent of plantation ( ha.) _______

e. Age of plantation________

f. Previous land use ____________

g. Topography ____________

**A. Biodiversity assessment**

**Plantation Other landuse Adjoining natural forest**

(i) Number of species

(ii) Relative dominance

(iii) Relative frequency

(iv) Relative density

(v) IVI

(vi) Shannon-wiener diversity index

(vii) Faunal diversity

**B. Soil assessment**

(i) Structure of soil

(ii) Nutrient status

(iii) Soil erosion

(iv) pH

(v) Soil compaction

(vi) Nutrient cycling

**C. Socio-economic factors**

(i) Employment status

(ii) Fulfillment of day to day requirements, viz. fodder, fuel, timber

(iii) Status of minor forest produces (medicinal plants, fruit etc)

(iv) Religious sentements

**D. Economics of plantation**

(i) Net returns (Rs/ha)

(ii) Average annual income

(iii) Cost of establishment and maintenance

* After obtaining these informations the comparison is made between plantation and other land use by converting the observation into decimal fraction. Eg. (A) If number of species in plantation was 3 and in adjoining forest was 15 so 3/15 = 0.2. ⇒ plantations were having 1/5$^{th}$ species compared to forest. (B) If net return from plantations were Rs. 100,000/ha and from forest Rs 50,000/ha so 50,000/100,000 = 0.5. ⇒ natural forest gave ½ income compared to plantation.

** To avoid confusion, + and –ve sign is added to denote whether return from plantation was less or more compared to natural forest or other land use.

**Do it yourself**

1. Compare the floral diversity of plantation with that of natural forest.
2. Observe the effect of plantation on socio-economic status of people residing nearby.
3. What is the effect of plantation on soil characteristics.

Chapter **7**

# Energy and Plantation Forestry

Energy is the capacity to do work. All movements and activities in this world is governed by change in one form of energy into another. Rate at which energy demands and prices are increasing it may be impossible to pursue the present rate of development. Energy has always been a vital resource in the development of any nation. The prosperity of a nation is measured in terms of per capita energy consumption besides GDP, GNP, etc. While the world has seen hectic industrial activity in the past century it has also come face to face with serious problems arising out of haphazard utilization of the energy resources. Developing countries, like India, will be forced to retard its development/industrialization program for want of sufficient energy reserves. Share of energy production cost varies greatly from 1 per cent to 36.5 per cent. In industries like ferroalloys energy share is 36.5 per cent, cement – 34.9 per cent, pulp & paper – 22.8 per cent, iron and steel – 15.8 per cent, sugar – 3.4 per cent, refineries – 1 per cent. This implies that large amount of production cost has to be incurred on energy. In developing countries where energy supply is irregular and insufficient, this share may even go higher. Large number of form of energy exists; they are broadly categorized into two types viz. renewable and non-renewable. The environmental risks of fossil based or nuclear fuel cast doubts on their future viability as stable energy sources. Knowing the fact that non-renewable energy sources are finite and are associated with environmental pollution, strategies to switch over to renewable energy sources are gearing importance. After all a finite planet cannot support infinitely increasing consumption of resources and hence the motto of present times must be to "REDUCE, REUSE, RECYCLE".

**Promises of renewable energy**

— They can produce power at costs that are competitive with conventional energy sources.

— Compared to conventional power generating system renewable energy systems are viable for rural and remote areas, as resources for generating it are locally available eg biomass, wind and sunlight.

— It has environmental benefits.

## IMPORTANT RENEWABLE ENERGY SOURCES

**(a) Hydroelectric power**

Electricity produced through hydroelectric power is though renewable but for establishment of hydroelectric power setup, large extent of forest lands are to be sacrificed and it also disrupts the ecology of the river fauna. Moreover, construction of large dams is costly and hence they cannot be constructed at frequent distances. Thus scarcity of electricity is often being seen. This scarcity will significantly impair the economic growth of developing countries in next four decades. In many developing countries electric supply systems are falling into despair because of poor management and lack of investment in maintenance- frequent black outs and poor quality power. Sanghvi (1991) has established that insufficient and unreliable power can exceed $ 1.00/kwh leading to loss of wealth. Rural communities and rural industries have to cope with the least reliable and lowest quality power supply that is typical at the end of transmission line.

### (b) Solar energy

Sun is the ultimate source for all forms of energy. Sun, is over a hundred times larger than the Earth in size and is located at a distance of 150 million kilometer from the Earth. It is said that Sun is a fusion reactor emitting 3,800 million, million, million, million watts of energy each second and the earth receives only 1/2,00,000,000,000 portion of this amount which is equal to 1.3 X 10^17 w/h. which is 20,000 times the energy requirement of the world. About 50% of the energy received outside Earth's atmosphere actually reach the Earth and is about 1 cal/sq.cm. min at sea-level. The solar energy can be broadly classified in two categories on the basis of its use – (A) Solar Active (Direct Use) and (B) Solar Passive (Indirect Use).

### Solar Active

In Solar Active category the solar energy is directly converted in the application form and can be further divided into two forms – Solar Thermal (Heating Application) and Solar Photovoltaic (Electricity Generation).

#### *(i) Solar Thermal*

Solar Thermal Technology is employed for collecting and converting the sun energy to heat energy for application such as water and air heating, cooking and drying, steam generation, distillation, etc.

### *(ii) Solar Photovoltaic*

Solar Photovoltaic Technology is employed for directly converting solar energy to electrical energy by the using "solar silicon cell".

### Solar Passive

In Solar Passive, the solar energy can be put into use by incorporating appropriate designs in buildings itself to maximize utilization of solar energy for various purposes such as lighting, seasonal air conditioning, water and space heating/cooling and thereby reduce external energy inputs.

Despite number of advantages and utilization, solar energy cannot be tapped throughout the country because of variation in sunshine, due to altitude, fog, cloudy weather and rainy days.

### (c) Wind power

Wind is global phenomenon occurring on the earth's surface due to unequal heating of various parts of the earth's surface by the sun. The power contained in the wind can be harnessed for mechanical application such as water pumping or for generation of electricity, through wind turbine generators. Wind is affected by the terrain and by height above the ground. Wind speed generally increases with the height above the ground as moving wind moving across the earth's surface encounters friction caused by the turbulent flow over and around the mountains and hills. Also due to heating and cooling of water and land, wind movement is more in costal areas. However, wind speed and direction vary in the short and long term. There is a variation in minute to minute. Thus this energy has its potential only in some specified localities only.

### (d) Tidal energy

Tides are generated through a combination of forces exerted by the gravitational pull of the sun and the moon and the rotation of the earth. The relative motion of these three bodies produces different tidal cycles which affect the range of the tides. Energy can be extracted from tides by creating a reservoir or basin behind a barrage and then passing tidal waters through turbines in the barrage to generate electricity. Since India is surrounded by sea on three sides, its potential to harness tidal energy has been recognized by the Government of India. Potential sites for tidal power development have already been located. The most attractive locations are the Gulf of Cambay and the Culf of Kachchh on the west coast where the maximum tidal range is 11 m and 8 m with average tidal range of 6.77 m and 5.23 m respectively. The Ganges Delta in the Sunderbans in West Bengal also has good locations for small-scale tidal power development. However, Tidal energy is extremely site specific and requires mean tidal differences greater than 4 meters and also favorable topographical conditions, such as estuaries or certain types of bays in order to bring down costs of dams.

## (e) Geothermal energy

Geothermal energy is the natural heat generated from within the earth. Underneath the earth's relatively thin crust, temperatures range from 1000 to 4000$^{0}$ C and in some areas, pressure exceeds 20,000 pounds per square inch (Psl). Geothermal energy is most likely generated from radioactive thorium, potassium and uranium dispersed evenly throughout the earth's interior which produce heat as part of the decaying process. About 10 percent of the world's land mass contains accessible geothermal resources and could provide several million quads of energy annually. Geothermal energy is derived using the earth's natural fluids to extract thermal energy that can be sent through turbines to generate electricity or be used directly e.g. direct heating and industrial process heat. Geothermal resources are widespread in India, in the form of 340 hot springs sites.

As it is uneconomical to transport the thermal energy over long distances, the end use is often located near the source.

## BIOMASS AS A SOURCE OF ENERGY

Biomass has been used as the source of energy since the advent of fire. However, with the availability of other energy sources its use got decreased. Looking into the account the increased pollution by use of fossil fuel and some of the problems associated with other renewable energy sources, use of biomass as source of energy is again getting importance.

### Advantages of biomass

— Combustion of most types of biomass feed stocks emits little or no sulphur and less nitrogen oxide then burning fossil fuel. Hence there will be reduction in pollutants causing acid rain and smog thus improving the air quality. An additional environmental benefit is in water qualily, as energy crop fuel contains less mercury then coal (Pace university, 1990). Biomass energy combustion results in no net emission of $CO_2$. It doesn't emit methane when burnt, otherwise it emits methane if it is let to decay naturally in land fills (Klass, 1991).

— Production of biomass can rehabilitate degraded lands, use of biomass can solve waste management problem

— Because biomass / bioenergy production operations are decentralized and labour intensive, rural employment can be another economic benefit.

— Alive and active growing forest and other plant biomass absorb the greenhouse gas (Co,) in quantities broadly equivalent to the amounts emitted when plant materials decay or are burned. They are thus called as "Carbon-neutral" fuel sources.

— Since trees absorb and store atmospheric carbon as they grow, which is known as carbon sequestration, growing plantation forest will reduce level of $CO_2$. Biomass energy production reduces green house gases in two ways (a) Biomass energy crops are "carbon cycle neutral". (b) Growing energy crops creates a "carbon sink" which includes storing carbon underground through the tree root system. Because of this creation of "carbon sink" (a component which solar and wind energy don't have) we believe that closed loop biomass energy crops represents the most effective choice in renewable energy option to address global warming.

— Biomass energy production have lower energy cost per unit area as lower inputs are requires with regards to fertilizers and pesticides compared to annual crops.

— Upliftment of rural economy and local employment generation are associated with biomass energy production particularly in form of managing plantation. It will thus check urban migration.

— It is also feasible in small-scale production system in form of farm forestry where industries will supply seedlings, fertilizers and technical know how to farmers and also ensure purchase of harvest.

— Energy plantations are thought to remove the entire nutrient from the soil. However, by use of thermo-chemical process of biomass conversion it is feasible to recover all mineral nutrients as ash which can be returned to the plantation sites thus there will be no loss of nutrient from the soil.

## COMPARISON OF BIOMASS ENERGY WITH OTHER SOURCES

Estimate of environmental cost of electricity production from different sources were made with reference to biomass.

It was found that biomass pollute the environment several times lesser then fossil fuel in generating same quantity of electricity (Table 1) (Pace university, 1990)

Table 1. Comparison of different sources of energy with regards to pollution potential.

| Source | Minimum pollution | Maximum pollution |
|---|---|---|
| Solar PV | 0 | 0.4 |
| Wind | 0 | 0.1 |
| Biomass | 0 | 0.7 |
| Coal | 2.8 | 6.8 |
| Oil | 3.0 | 7.9 |
| Natural gas | 0.8 | 1.2 |
| Nuclear | 2.91 | ? |

It had been estimated that poplar coppice can provide about 6 tonnes of oil equivalent per hactare compared to only 1 tonnes of oil equivalent per hactare for rape seed oil. If all suitable farmland in Britain were planted with rape seed it would provide only a diesel substitute equivalent to about 2 per cent of national liquid fuel requirements, whereas, just 1 million hectares of wood fuel plantation could meet 10 per cent of U.K solid fuel needs.

## From where wood will come

As it is known that biomass is the only fuel which can sustain in future but it is required in huge quantity, the question is from where we can meet these huge demands. We have to look for the areas where wood is being wasted while being processed. Table 2 presents percent wastage of wood in different wood industries, which could become a huge source for biomass energy.

Table 2. Per cent wastage of wood in different wood industries

| Source | Waste (% of unprocessed log) |
|---|---|
| Saw mills | 55 |
| Veneer and plywood | 55 |
| Fiber and particle board | 20 |
| Pulp and paper | 20 |

Asia has largest volume flow through its wood markets and as a consequence has the largest potential to produce energy from wood residue. In Asia, India has largest potential followed by Indonesia in providing wood residues for electricity production. To help ensure minimum competition between agriculture and energy plantation number of analysts have proposed that developing countries target degraded lands for energy production. It has been estimated that developing countries have over 2,000 million hectares of degraded lands which can be utilized for energy plantations. However, restoring these lands is a costly affair. Energy industries requiring raw material might provide the capital to farmers/ owner of these degraded lands for restoring them. India can become major bioenergy producer without compromising food production needs. It had been reported that the net cropped area, of India, between 1970-1990 remained almost same but population increased 2.4 times still India achieved self sufficiency in food. However, till date production per unit area of all agriculture is quite less then other Asia countries. Eg. India's average rice production is 1.7 tonnes/ha, which is about half the Asian average. Thus food production can still be doubled without increasing the agriculture area. As a result the degraded lands available can be used for biomass energy production which are available in a range of 80 –1 00 million hectare. If only 65 million hactare of these degraded lands are considered for biomass energy production with average dry biomass yield of 10 tonnes/hac/year and a biomass energy content of 20 GJ/dry tonne, 13 EJ of energy could be produced which is more then the total primary energy consumed in India (11 EJ) in 1991 (World Resource Institute, 1994).

## PROBLEMS ASSOCIATED WITH ENERGY PLANTATION

— For producing large biomass per unit area, high dozes of fertilizers, herbicides, pesticides and intensity of water may lead to pollution.

— These large-scale plantations may lead to decrease in local biodiversity.

— In many developing countries resource conflicts and constraints can spoil or complicate sustainable production and use of wood fuels. Demand of fuel in cities can affect rural livelihoods and environmental stability, from where wood fuel will be harvested, due to over exploitation.

— Disputed land ownership, lack of roads to transport biomass to processing industries and biofuel to market, long wait for harvest are some the endemic problems associated with popularizing plantation/ biomass/wood fuel energy.

**How can biomass energy be encouraged ?**

— Removal of subsidies for fossil fuel will encourage reconsideration of investment in timber /biomass production.

— Creating energy literacy among citizen so as to enable them to make wise energy choices to promote and popularize renewable biomass energy sources and energy conservation measure.

— Government can take leading role by preparing and implementing policies related to biomass energy production as is done in Gujarat where "Wind power generation policy-2002" had been formulated.

## FUTURE STRATEGICS

The use of fuels derived from forest and other biomass has tremendous potential as an environmentally friendly, technically mature, economically feasible and socially acceptable energy practice. To popularize it the need is to develop:

a. New and improved wood energy conversion technologies to produce more energy using less wood.
b. Better flow of wood fuel between producers and users.
c. Increased fuelwood production at lower prices.
d. Adequate and good quality supplies of wood fuel need to be readily available throughout the year.
e. Advancement in fuel processing techniques like fuel briquettes from plant waste need to be emphasized.
f. Developing technologies for densification of wood, which will have, even consistency for rapid conversion to energy products without prolonged drying and other preparation.
g. Principles of sustainable agriculture have been recommended for biomass plantations to maintain productivity and benefit the rural economy while minimizing adverse impacts to soil, water quality and wild life.
h. Harvesting of plantation crop must be timed to protect important wildlife species and to reduce soil erosion and non-point pollution.
i. Developing wooden stoves for rural houses, which are efficient with regards to energy release and less polluting.

**Future Prospects of biomass**

Biomass today accounts for 1/3rd of all energy used in developing countries. It had been called as "Poor man's oil" because of its direct use for cooking and heating. Existing biomass using technologies are relatively inefficient thus biomass provides less energy service then the proportion of total energy it represents. Most biomass energy today comes from natural forests, contributing to deforestation in many countries. UN Conference on Environment and Development (UNCED) found that sustainable biomass energy system could be largest single contributor to global energy supply. The study under "Renewable – Intensive Global Energy Scenario (RIGES), biomass could provide as much as 35 per cent of total demand for primary energy in 2050 (Johansson *et al.*, 1993).

## DIFFERENT TYPES OF CHULAHS AND STOVES

At present and in future the major use to which biomass is being or will be put to is in the house holds for cooking and heating purposes. Livelihood patterns in rural India are characterized by biomass-based systems. Cooking is the largest energy consuming end-use in the domestic sector, accounting for almost 90% of the total domestic energy use. Wood, crop residue, and animal dung are the main fuels available and used in rural areas. Thus there is a need to give emphasis on improving the design and efficiency of the stoves and chulahs which are presently being used in rural areas so that with use of less biomass and less pollution more energy from same amount of wood can be obtained. The present level of fuelwood consumption is around 200 million tonnes per annum (TERI, 1997), while it is estimated that the annual sustainable yield of fuelwood is only about 86 million tonnes per annum. Thus, there is a wide gap between demand and sustainable supply. The inefficient use of biomass in traditional devices has serious environmental effects and leads to high levels of indoor air pollution, which affects women and children during cooking.

## Designs of cooking stoves and chulahs

There are various designs of improved chulahs, which are available. The local and unimproved chulah is given in photo. 1. The choice of kind of chulah depends upon the cost, working condition, kind of fuel used etc. Following are the types of chulahs that are common in use.

**Astra chulah :** ASTRA stands for Centre for "Application of Science and Technology for Rural Areas", IISc which has been looking at the rural cooking problem and trying to identify solutions - to improve the efficiency of cooking stoves. To improve the design of cooking stoves ASTRA started with the following assumptions:

i. Any new design should be based on the geometry of existing chulahs;

ii. The stove should be made mostly from locally available materials by local people with minimum additional skills and with minimum training;

iii. The stove should accept local fuels - twigs, bark, husk, bagasse, stems etc;

iv. It should not cost more and it should not put any constraint on existing practices.

With the above conditions, the objectives chosen were;

(i) Ease of cooking; (ii) Minimization of smoke; (iii) Ease of starting and stopping the fire; (iv) Maximization of efficiency.

The stove designed is having three pan, mostly made of mud-looking similar in appearance to existing stoves. There is an enclosed fuel box with a cover; the grate, and parts for allowing primary air into the chulahs; flow of fuel gas is controlled by the ducts between pans (1 and 2, 2 and 3) and chimney. The pan openings can be adjusted during construction so as to fit the vessels used normally in the house. To achieve the given objectives the following were modified or incorporated; (a) area of primary air inlet hole; (b) distances (heights) of combustion chambers; (c) flow diffuser for third pan; (d) orientation of fuel box; (e) materials used for construction and insulation of inner wall-mud and rice husk; (f) coating the grate with simple catalysts to facilitate cracking. A number of models have been constructed and tested before arriving at an optimal model, ready for diffusion amongst villages. The production model named Astra ole was taken up for field tests - in the high rainfall Uttara Kannada district and the low rainfall Tumkur district. Results were encouraging. ASTRA put up about 100 stoves in Ungra and neighbouring villages. Field efficiencies were varied from 30% to 43% with a mean of 35.5%. This is nearly double that of a traditional stove (mean is 13.3%). Similarly, specific fuel consumption varied from 100 to 250 grams with a mean of 168 grams. The comparable figures for traditional stoves are 251 to 450 grams with a mean of 351 grams. Hence it is clearly shown that the people can save definitely half of the firewood used previously. Astra ole is smokeless (so cooking is a pleasant activity for village women); it cooks faster - 40% lesser time for cooking than that of conventional ones, as flame is steady and continuous, no blowing of air is required - improves health of women doing cooking activities; various types of fuels can be used with ease.

**Priagni**

This is an improved wood burning stove designed by the Rural Energy Laboratory, Central Power Research Institute. It has been designed to improve efficiency, to create a steady and continuous flame for burning and to reduce smoke. It is an iron single pot stove modification of existing commercially available ones. The factors considered in design improvement are: (i) Proper height of the stove (combustion chamber) - low height means incomplete combustion and high height means more losses to atmosphere; (ii) Optimal diameter - to reduce smoke; (iii) Height of vessel mounts; (iv) Fuel feed opening - to reduce exposure of wood flames (thereby reducing radiation loss); (v) Grate design.

After optimisation of these parameters, two components have been added. The first is a set of two slotted plates - one at the top and the other at the bottom. The top plate creates turbulance, imparts velocity to the gasses and acts as a partial enclosure to combustion chamber. The bottom plate acts as an air inlet providing for symmetric air flow - helps in continuous burning of the fuel. The second addition is a clip - on aluminum lining to the inner wall. The thin air gap between the stoves wall and the lining reduces heat flow to the walls, thereby increasing efficiency. The costs vary from Rs. 35/- to Rs. 78/-. Because of its resemblance to existing stoves, it is easier to operate and so acceptability particularly amongst urban households is high.

**Improved Metal (imet)** : This is a portable metal non-chimney woodstove with a single pothole developed in 1983 by Central Power Research Institute (CPRI), Bangalore, India. In 1991, the stove was brought under Indian standards (BIS 1991).

**Improved Vented Mud (ivm)**: This is a two-pot cookstove with chimney, called the *Nada chulha*. A tunnel connects the fire box to the second pot hole and to a chimney. Since two pot holes are provided two things can be cooked on it at the same time with only one fire.

**Improved Vented Ceramic (ivc)**: This is also a two-pot cookstove with chimney. Made of a ceramic lining with mud coating, this stove was developed at the Central Glass and Ceramic Research Institute, Khirja, Uttar Pradesh, which is one of the Technical Back-up Units of the national improved stove program.

**Hara**: This is a traditionally designed earthen pot for burning dung cakes and used mainly for slow heating of milk over three to four hours such that, without boiling, the cream of the milk separates as a thick layer at the surface. It is also used for cooking fodder.

**Angethi (used for charcoal and charbriquette) :** This is a portable stove fabricated with a galvanized iron bucket, mud/concrete, and grate. The fuel has to be fed above the grate by lifting the pot in a batch operation.

## Sampatharajan Biomass Gas Stove

This biomass gas stove works on the principle of updraft gasification. Biomass such as woodchips, agricultural residues, dry leaves, barks can be used in biomass stove. The stove is cylindrical drum made of clay, sand and paddy husk. It is 290 mm in diameter and 630 mm in height. An iron gate is fixed at 50 mm from the base to hold the biomass. The openings for ash removal and secondary air supply are provided (Photo 2). To operate the stove a head of ignited charcoal (1/2 kg to a depth of 50 mm) is placed over the grate. Then the biomass is dumped over red hot charcoal. A white gas emanates and it is ignited. Experiments shows that stove consumes 3.4-4.5 kg of biomass per hour. As the material is gasified and burn at higher flame temperature (600-700°C) an increase in thermal efficiency (23-25%) is achieved, heat output is 4.4 kw and saving in fuel over the conventional wood stoves is 10% and saving in times is 40%.

**Cook stove with imbedded water tank :** This improved chulha has a sliding firebox door and chimney dampers to control air supply for combustion. Apart from fuel saving, the main advantage of the stove lies in hot water availability, almost throughout the day, as the embedded water tanks also act as hot water storage devices.

## Case study of improved cook stove in Himachal Pradesh

The rural energy consumption pattern depends on geographical, cultural, social, and economic factors. The energy consumption pattern for energy and fuel-mix also vary considerably across different agro-climatic zones. For example, firewood, crop residue and dung cake are consumed in the Western Himalayan zone at the rate of 1.95 kg/cap/d, 0.14 kg/cap/d, and 0.22 kg/cap/d, respectively; whereas in the East Coast Plains and Hills, the consumption rate of these bio-fuels is 0.60 kg/cap/d, 0.08 kg/cap/d, and 0.13 kg/cap/d, respectively (Planning Commission, 1992). The kind of bio-fuel, its amount and availability on a daily basis, and other factors that are specific to certain agro-climatic zones play a key role in establishing the energy consumption pattern. While a significant proportion of energy may be used for heating water, and cooking in cold areas such as the Western Himalayan zone, in the East Coast Plains and Hills, energy may only be consumed during cooking.

The bio-fuel consumption in hilly areas is high. Bio-fuel consumption in Himachal Pradesh is the highest (firewood - 1.31 kg/cap/d, crop residue – 0.14 kg/cap/d, dungcake – 0.22 kg/ cap/d) when compared to other states in India (NCAER. 1985), the environmental effects of indoor air pollution are also higher in the state. Therefore, it was thought to introduce efficient cooking stoves in these areas. TERI (Tata Energy Research Institute) took more then 20 villages in Kandaghat block of Solan District of Himachal Pradesh to design, introduce and test the improved chulah. The financial aid for which was provided by Indian oil Corporation Ltd and the Gas Authority of India Ltd.

**Approach:** In the study area, the main device used for cooking and water heating is the traditional three-pot mud cookstove. These stoves have low thermal efficiencies for cooking ranging from 8% to 10%, have a high thermal mass, are made of stone and mud, and are generally not provided with chimneys. Chir (*Pinus roxburghii)*, Banj (*Quercus*), and other shrubs are the common species used as fuel. The traditional cookstove was modified to address the need for hot water expressed by the local community. The design of the improved cookstove was developed with inputs and suggestion from users.

**Design** : Two version of improved chulah - with or without water tank (Photo 3) was developed. The improved cook stove designed for this region consists of two water tanks embedded in the stove structure. The stove was provided with a chimney, which sucked air for combustion and also removed smoke, and a front damper to maintain the temperature in the stove and water pipes. The tanks were connected to an overhead water tank, which supplied cold water. Hot water is drawn from a water tap provided in the kitchen or in the bathrooms.

## Results

It was observed that households saved fuel (Table 3) and cooking time with the improved model. The cost of the improved stove along with pipelines, fitting, and labour comes to around Rs 600-700 per system, excluding the cold water tank. This resulted in large scale acceptance of this improved stove and this techniques has also been extended to other villages of District Shimla of Himachal Pradesh.

Table 3. Comparison of traditional and improved stove for energy consumption

| House hold (HH) | Adult energy consumption/day | |
|---|---|---|
| | Traditional stove | Improved stove |
| HH1 | 57.84 | 22.15 |
| HH2 | 45.22 | 24.14 |
| HH3 | 37.9 | 25.84 |
| HH4 | 65.79 | 34.45 |
| HH5 | 60.3 | 38.76 |

## DETERMINATION OF MOISTURE CONTENT, CALORIFIC VALUE AND ASH CONTENT OF FUEL WOODS

While comparing different fuels for their efficiency with regards to energy they contain, observation on calorific value, moisture content and ash content are done. Following are the methods of determining these parameters.

### (A) Moisture content

Moisture content can be measured in two ways:

(a) on a dry basis ($M_{db}$), that is:

$$\frac{\text{Wet weight} - \text{dry weight}}{\text{dry weight}} \times 100$$

(b) on a wet basis ($M_{wb}$), that is:

$$\frac{\text{Wet weight} - \text{dry weight}}{\text{wet weight}} \times 100$$

Thus $M_{db}$ is always greater than or equal to $M_{wb}$ and a moisture content of 100% or more is possible when calculating on a dry basis. In some countries fuelwood moisture content is given on a wet basis. The formula for changing from dry to wet basis is as follows:

$$\frac{D}{1 + D/100} = W$$

where D = moisture content as a percent of the dry weight basis; and

W = moisture content as a percent of the wet weight basis

If the moisture content is 100 per cent on the dry basis then the wet basis moisture content = 50 per cent.

Similarly a 15 per cent dry basis moisture content = 13 per cent (wet basis).

The reverse formula for changing from wet basis moisture content to dry basis moisture content is as follows:

$$\frac{W}{1 - W/100} = D$$

* The moisture content of wood is obtained by drying wood in an oven at approximately 100 ± 3°C until its weight no longer changes. The quantity of moisture in the wood is determined by weighing the wood before and after drying.

**Example:** Find out moisture content of a wood, on dry weight and wet weight basis, whose weight at the time of felling was 50 kg. On drying the wood using standard method, the oven dry weight of wood became 29 kg.

**On dry weight basis ($M_{db}$)**

$$\text{Moisture content (\%)} = \frac{\text{Wet weight} - \text{dry weight}}{\text{dry weight}} \times 100$$

$$= \frac{50 - 29}{29} \times 100 = 72.41\%$$

**On wet weight basis ($M_{wb}$)**

$$\text{Moisture content (\%)} = \frac{\text{Wet weight} - \text{dry weight}}{\text{wet weight}} \times 100$$

$$= \frac{50 - 29}{50} \times 100 = 42\%$$

**Conversion of moisture content of wet basis to dry basis**

$$D = \frac{W}{1 - w/100} = \frac{42}{1 - 42/100} = 72.41\%$$

**Conversion of moisture content of dry basis to wet basis**

$$W = \frac{D}{1 + D/100} = \frac{72.41}{1 + 72.41/100} = 41.99\%$$

**(B) Calorific value**

While selling the wood for fuel purposes, weight is generally regarded as the best suitable method. However, weight varies with moisture content and density of wood. The density of wood varies with species (non-conifers usually have more density then conifers), with age (Juvenile wood in young trees is less dense than mature wood in old trees), sapwood is less dense than heartwood etc. However, what is important to note is that if the moisture content is the same, the energy given off from a piece of wood is more or less the same on a weight basis irrespective of species, and this is an important factor when planning the needs of a household, community or industry. Therefore, knowing the amount of energy stored in a piece of wood is a good way of assessing the fuel wood and hence it's selling and buying. Other advantages of converting a measurement of fuel consumption into units of energy are : (a) conversion to energy units corrects for differences in fuelwood type and quality. (b) conversion to energy units is necessary in order to compare fuelwood consumption with other energy demands. In addition, a quantitative estimate of cooking energy demand is needed to evaluate the potential for alternative fuel sources and new cooking technologies. Fuelwood consumption can be converted into units of energy consumption by multiplying it by the calorific value of the fuel (MJ/kg).

The two most important determinants of wood's energy content (calorific value - CV) are its density and moisture content.

The energy content of wood containing no moisture is referred to as the *Gross Calorific Value* [GCV]. When wood contains moisture, some of the energy will be used to evaporate that water when the wood is burnt and the net amount of energy that will be released is known as the *Net Calorific Value* [NCV]. The GCV of the wood of nearly all trees lies between 19 and 20.5 gigajoules *per* tonne [GJ/t]. A good, working figure for most trees that are likely to be grown in plantations is 19.5 GJ/t. This may also be expressed as 19.5 MJ/Kg.

The calorific value of solid fuels is normally measured using a bomb calorimeter (Bialy, 1979b). Bomb calorimetry yields the higher heating value (HHV), also known as the gross calorific value. A bomb calorimeter and other instruments needed for measuring calorific value cost about $1 000 (Rs. 47,000 approx.). Furthermore, these devices may be difficult to obtain in a developing country and some scientific training is required for their operations. In addition, the bomb calorimeter and related instruments are somewhat delicate and they must be used in a laboratory. For these reasons, it may not be possible to directly measure fuel calorific value during a fuelwood survey. As an alternative to bomb calorimetry, the calorific value of biomass fuels can be estimated with minimal error based on their moisture content and the known calorific value of dry fuels. Table 4 shows the approximate calorific value of different biomass fuels on an oven-dry basis. The oven-dry calorific value is relatively constant for each type of fuel. There is a small difference between the calorific value of hard and soft wood; the value for soft wood is approximately 5% greater because of its higher tar and resin content (Arola, 1978). In practice, the calorific value of fuel will differ from the values shown in table 4, mainly because of moisture content (Arola, 1978; Bialy 1979b). Moreover, this effect can be significant since moisture content varies over a wide range.

Table 4. Calorific Value (HHV) of Different Biomass Fuels on an Over-dry Basis

| Fuel | Calorific value (MJ/kg) ($E_{od}$) |
|---|---|
| Wood | 20 |
| Wood charcoal | 31 |
| Rice husks | 15 |
| Cow dung cakes | 18 |

The actual calorific value of fuel wood $E_w$ (HHV) can be estimated using the following equation, if the moisture content of the fuel is known.

$$E_w = (1 + M_{db}) - 1_{Eod}$$

where $M_{db}$ is the moisture content on a dry basis and $E_{od}$ is the calorific value of the fuel when oven dried (Table 4).

**Example:** Find out actual calorific value of fuelwood if moisture content of wood is 30 % on dry weight basis.

Calorific value ($E_w$) = (1+ $M_{db}$) – 1 $_{Eod}$

= (1+30) – 1(20) = 31 – 20

= 11 MJ/kg

This energy stored in wood can also be calculated with the following formula :

$$\frac{100x - 2.4\,(54 + m)}{100 + m}\ \text{MJ/kg}$$

where x = (high) heat value of wood ~20.0 MJ/kg. and

m = the moisture content. The factor of 2.4 MJ per kg of water is the amount of heat required to drive off 1 kg of water from the wood.

This formula shows that it is the moisture content which affects the energy of the wood. Eg. One kilogram of wood at 6.15 percent moisture content when burnt will give off about 16.0 megajoules ($16 \times 10^6$ J) (MJ/kg) that is 3,820 kilocalories (kcals) or 15,170 British thermal units (Btus). Wood with 40 percent m.c. has a heat value of 12.7 MJ/kg and that of a 100 percent m.c. 8. 2 MJ/kg.

**Example**: Calculate energy stored in wood if its moisture content is 30% on dry weight basis with high heat value of wood approximately 20 MJ/kg

$$\frac{100x - 2.4\,(54 + m)}{100 + m}\ \text{MJ/kg}$$

$$= \frac{100 \times 20 - 2.4\,(54 + 30)}{100 + 30} = \frac{2000 - 201.6}{130}$$

$$= 13.83 \text{ MJ/kg}$$

**Ash content**

In order to find the ash content of wood, 3 to 5 samples are taken from the dried samples. The chips are burned in a steel container (muffle furnace) at 550°C for eight hours. If no carbon or wood is found in the sample, the oven is switched off. If wood or carbon is found, the sample are checked every two hours until it had burned completely. The samples are removed once the temperature of the oven decreases to 100°C. After having cooled down, the samples are weighed and their ash contents calculated as the proportion of ash of the dry weight of the sample (Asikainen and Pulkkinen, 2002).

## ENERGY CONSUMPTION SURVEY IN RURAL/URBAN AREAS

(a) Village name________________

(b) Name of Head of family________________

(c) Number of family members________________

(d) Annual income ________________

(e) Category of family: Small farmer/medium farmer/ large farmer/agricultural labour

(f) Sources of energy used : Wood/kerosene/ LPG/ coal/ electricity/crop residues

(g) Species used for burning________________

(h) Source of wood: Collected from near by forest/ purchased from market

(i) Amount of wood used for cooking/ day________________

(j) Season of collection of wood and its storage method________

(k) Time consumed in cooking________

(l) Problems faced while cooking (pollution etc)________

(m) Type of chullah used for cooking________

(n) Moisture content of wood ________

(o) Actual calorific value of wood ________

(p) Total calorific value used per day (g x m)________

(q) Amount incurred in cooking (with reference to fuel) per day (g x rate of fuel) ________

(r) Ash content generated per day while cooking (ash content/gram of wood x g)________

(s) Where the ash is disposed________

* The calorific value of kerosene, LPG etc are readily available.

## Calculation of fuel wood energy balance for a region

Planning for meeting fuel wood energy requirement of a region is important. In planning, knowing energy requirement and sources from where it could be met is desired. Following is a method of estimating energy balance of a region.

## Format for estimating Fuel wood energy balance

1. Name of the District/town/area________

Total area ________

Agriculture land ________

Land under forest ________

Degraded land/wastelands which can be used for raising energy plantation ________

Steep mountain, lakes, river and urban area ________

Population________

Rural ____________ Urban ____________

Rate of population growth ____________________

Annual per capita rate of fuel wood use (kg/year or $m^3$/year):

Rural ____________ Urban ____________

Volume/weight of fuel wood exported from region ________

Volume/weight of fuel wood imported to region___________

Volume of wood extracted from present forest ___________

**Example**

Name of the District/town/area "XYZ"

Total area : 50 ha

Agriculture land: 20 ha

Land under forest : 6 ha

Degraded land/wastelands which can be used for raising energy plantation: 6 ha

Steep mountain, lakes, river and urban area: 18 ha

Population: 7 lakhs

Rural: 2 lakhs Urban : 5 lakhs

Rate of population growth : 1 %

Annual per capita rate of fuel wood use (kg/year or $m^3$/year):

Rural: 5kg/day/person (1825 kg/person/year) $\Rightarrow$ 3.65 lakh tones/year

Urban : 2 kg/day/person (730 kg/person/year) $\Rightarrow$ 3.65 lakh tones/year

Total requirement: 7.30 lakh tones/year

Volume/weight of fuel wood exported from region : 2 lakh tones/year

Volume/weight of fuel wood imported to region: 0.5 lakh tones/year

Volume of wood extracted from present forest : 3 lakh tones/year

⇒ Demand of fuel wood = 7.30 lakh tones/year

Actual supply of fuel wood = 1.50 tones/year (extracted – exported + imported i.e 3 – 2 + 0.5).

In order to bridge this gap between demand and supply (7.30 – 1.50 = 5.80) of fuel wood one of the following or combination of these can be done:

1. The degraded/wasteland of the region should be planted with fuel wood species.
2. Reduce the fuel wood consumption by introducing improved cook stoves/ chulahs.
3. To meet out domestic requirements export of fuel wood from region can be decreased.
4. Other renewable sources of energy viz. solar, wind etc should be encouraged to meet overall requirement of energy of the region.

**Do it yourself**

1. Calculate calorific value of important fuel wood of your locality
2. Find out the ash content of important fuel wood species of your area.
3. Survey a village, which is far from city and a village near city and compare them with regards to fuel wood consumption and variation in woodstoves.

Chapter **8**

# Plantation Protection and Economics

## Protection of Plantation

Protection is a very important and necessary component in man made plantation planning and management. It is very expensive operation. Plantation requires protection against grazing by domestic, migratory and wild animals. Protection against grazing and browsing activities is utmost important as it can destroy the plantation altogether. Even the unpalatable hardy and thorny species get damaged due to trampling by the animals. Fencing is also necessary for delineation of the plantation by fixing the boundaries of plantation and for ascertaining the areas of the plantation. There are various kinds of fencing for protection of plantation from grazing depending upon availability of material and labour. The commonly used fencings are:

(i) Stone wall fencing (ii) Brick wall fencing (iii) Trench/ ditch fencing (iv) Live hedge fencing (v) wooden fencing (vi) Barbed wire fencing (vii) Social fencing.

Out of these mostly live hedges and barbed wire fencing are generally used.

### (A) Live hedge fencing/ Biofencing

A dense impassable line of live hedge of easily propagated species viz. *Acacia nilotica, Prosopis juliflora, Zizyphus jujuba, Rosa macrophylla, Rosa* spp., *Rubus* spp., *Carissa oppaca* etc may be used for protection of plantation. The species to be used should be thorny and bushy with spreading branching system thus forming a meshy growth. The species should also be quick growing and unpalatable. The shrubs which are unpalatable can be used as live hedge barriers eg *Adhatoda vasica, Vitex negundo, Calatropis procera, Cassia auriculata, Jatropha curcas, Sesbania aegyptiaca, Lawsonia alba, Duranta plumerii, Agave americana, Euphorbia royleana, Dodonia viscosa*.

Following is tabular representation of the amount of money that needs to be incurred for protection of one hactare through biofencing.

**Investment in protection of one hactare plantation area through biofencing**

| S. N. | Operation/ activities | Man days generated |
|---|---|---|
| 1 | Seed collection of hedges/ shrub species | 2.5 |
| 2 | Ditch digging (30 cm deep) along the boundary of plantation of one ha (100 x 100m) | 16 – 20 |
| 3 | Sowing of seeds in ditch | 1 – 2 |
| 4 | Digging of pits (45 $cm^3$) in hedge row 1 m apart (400 pits in 100 x 100 m) | 20 – 24 |
| 5 | Planting of entire plants (naked roots) in hedge row | 6 - 8 |
| 6 | Maintenance of hedge row for two years | 10 – 15 |
| | Total | 55 – 71.5 |

## (B) Barbed wire fencing

The most commonly and extensively used fencing in forestry is barbed wire fencing with fence posts of either cement concrete or angle iron or wooden posts. The cement and angle iron posts are very expensive as compared to wooden fence posts. The latest technique used in Himachal Pradesh by Katoch, 1994 and some modification by the scientists of University of Horticulture and Forestry, Solan is discussed here in detail along with its economics.

### Design of latest fencing technique

Fig. 33 shows the design of the three-stranded barbed wire fencing along with the spacing between the wires and method of erecting the fencing. Fig. 34 shows the design of four-strand barbed wire. In case of three-stranded barbed wire the vertical spacing among the strands is 15 cm between ground level and the first strand, 22.5 cm between first and second strand, 30 cm between the second strand and third strand. The fence post should be fixed firmly in hole of 15 cm x 15cm x 45 cm depth. In the earlier fencing methods, the linear spacing between wooden fence post use to be 3 m apart. Thus for 10 ha of plantation the number of posts works out to be 540-750 depending upon the geometry of the plantation/land. Larger the distance between the posts lesser will be the number of posts required. The average number of fence posts in old method works out to be 645. In case the shape of the plantation is a square, the number of fence posts in 10 ha plantation works out to be 421. Further more if the shape of 10 ha plantation is perfectly circular the number of fence post required will be around 374. But in field it is not possible to find perfectly circular and square shaped area for plantation. Therefore, number of fence post has to be standardized as per the geometry of the area. Accordingly for 10 ha plantation, 600 fence posts are considered as standard number. In the new fencing technique the wooden fence posts are kept at 6 m apart and one vertical strand of GI wire (3mm) is interwoven in between, thus interlocking all the horizontal barbed wire as given in Fig. 33 & 34. In the latest technique the maximum upper portion above the last barbed wire strand kept is 7.5 cm. This further helps in saving of wooded posts up to 10 per cent.

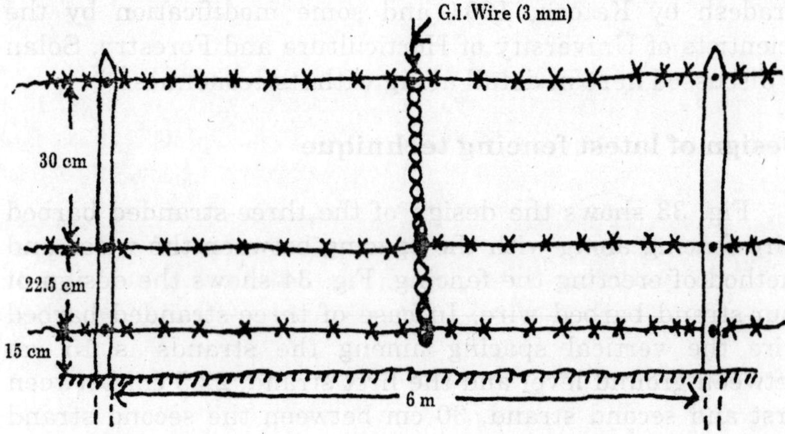


**Figure 33.** Design of vertical spacing three strand barbed wire

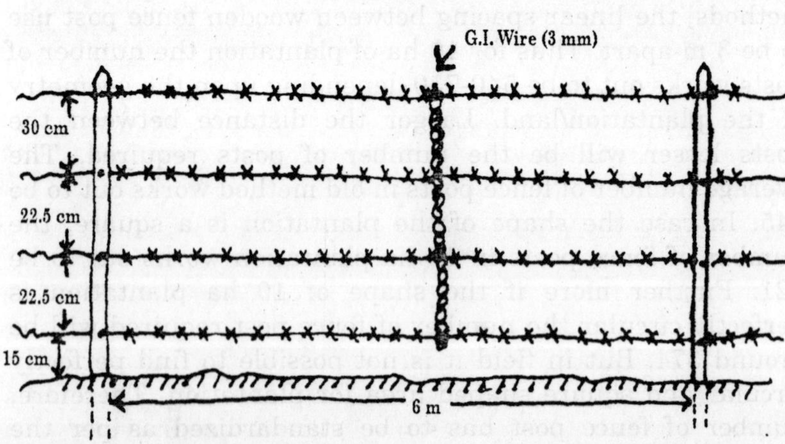


**Figure 34.** Design of vertical spacing four strand barbed wire.

**Cost of fencing of 10 ha plantation using wooden fence post of chirpine or eucalyptus or bamboo**

| S. No. | Particular/Activity | Material/labour cost (Rs) | Mandays | Total cost (Rs) |
|---|---|---|---|---|
| 1 | Cost of wooden fence for 10 ha of land (Approx. 600 post) | @ Rs 15 - 20 | - | 9000 – 12000 |
| 2 | Layout of the area for fencing | @ Rs 60/manday | 3 | 180 |
| 3 | Digging of pits 600 no's (15x15x45 cm) | 30 pits/man day | 20 | 1200 |
| 4 | Fixing of wooden fence posts in 600 holes/pits | 60 fence post/ man day | 10 | 600 |
| 5 | Carriage cost of 600 fence post to fencing site + loading and unloading | @ Rs 500 for 600 poles (tractor charge) | 4 | 740 |
| 6 | U strapples (30 g each) (2000 – 2500 strapples) | @ Rs 35/kg | - | 2000 |
| 7 | Cost of barbed wire of 11,000 – 14,500 m length i.e 3 or 4 strand barbed wire (1 kg = 7 m length) | @ Rs 35/kg | - | 55,000 – 73,000 |
| 8 | Fixing, weaving and carriage of barbed wire (Approx. 11000 – 14500 m in length) | @ 200 m/ manday | 55 - 73 | 3300 – 4380 |
| | Total cost of 10 ha. | | 92 - 110 | Rs 72,020 – 94,100 |
| | Cost on 1 ha. | | | Rs 7,200 – 9,400 |
| | Cost of per fence post | | | Rs 90 - 160 |

Katoch, 1994 had calculated cost of per fence post while fencing 8 ha plantation in Hamirpur range and 5 ha land in Barsar forest range of Himachal Pradesh which came out to be Rs 125-150 per fence. In present case the cost of fencing in one hactare came out to be Rs. 7,200- 9,400 depending upon shape of the land. The cost per fence post is about Rs 90 – 160. The latest technique replaces 50% wooden post by

G.I. wire strands. This replacement automatically reduces fire hazard, termites and fungal attack to the extent of 50% and subsequent fencing repairs will also be less. G.I wire strands will remain intact through out the closure period of plantation. Labour requirement is also much less then fixing wooden posts. Workability is very easy. It is less time consuming. It reduces sagging of barbed wire strands. Aesthetically, it gives very clean and neat look. This technique is also environmentally friendly as it minimize felling of trees for preparing post.

## (C) Social fencing

In previous section the cost of fencing has been calculated. It is evident that fencing cost is high. In cases where plantation programme has not been funded by Government agencies, it is difficult to put fencing around the plantation. In such areas, the plantations are on risk from grazing. The barbed wire fencing requires continuous maintenance from year to year and hence in short term as well as long term, it is costly. Live fencing certainly requires investment once, however, establishment of live fencing should be started 2-3 years prior to establishment of plantation. There are less species available which can be used as live fencing, having character of fast growth and are un-browsable. The chance of success/survival of live fencing species itself is questionable in all kinds of sites/soil conditions. Taking into account high cost of barbed wire fencing and long duration and problem in success of live fencing these fencing cannot be advocated in all conditions. An alternative to these problems is social fencing. Social fencing means "Protection of plantation by the society itself without using any physical barrier". It has also be defined as "Social fencing is a practice wherein people voluntarily protect forests without any enforcement" (TERI, 2005). Though social fencing is a cheap and effective method of protection of plantation but if it is not well coordinated, it may prove to be total failure.

**Steps in Organizing Social Fencing**

— Choose a site, which is approachable to all the stake holders.

— Supplement indigenous knowledge and the involvement of the whole village community in decision making

— Committee of responsible people representing all section of society of that area including women, should be formed

— Species to be planted should be chosen taking into account the preferences/need of the local people. Preferably multipurpose and short rotation species.

— Involve people in plantation program i.e. digging of pits, plantation and irrigation etc.

— Distribute equal benefits to all the stake holders.

— The villagers themselves should put some social penalty, on a person who will be caught damaging the plantation.

**Plantation Economics**

Plantation works are labour intensive and the cost of various operations varies from place to place and several factors e.g. species, payment of wages, method of plantation, age and type of planting stock, intensity of soil working and various maintenance operations. Hence the cost per unit of various operations in monetary terms will serve no purpose, however, it is pertinent to workout man hours or man days of work required for various purpose. A man day is equal to 8 man hours inclusive of a 20 – 30 minute lunch break in a day. Plantation cost consists of mainly three types of cost:

1. **Organization cost** : It includes the cost incurred on the smooth running of the organization eg office and its related cost.
2. **Labour cost**: It includes payment of wages
3. **Material cost**: It includes cost incurred on implements eg tractor with trolley, water tank, axes, pickaxes, spades, sickles, water pump, manures, fertilizers, rope, cost of seedlings and cost of chemicals.

The latest trend is to quantify the work done in terms of man days rather then in monitory terms. The expenditure can be calculated by multiplying the number of man days with the labour rate of the fiscal year. Man days generated for carrying out plantation per unit area have been worked out by Panwar and Bhardwaj, 2005 for man made plantation in Solan, Himachal Pradesh.

**Following is a general economics for carrying out plantation**

| Sr. No. | Operations/Activities | Quantity of work/ day | Total mandays (MD) |
|---|---|---|---|
| A | **Preplanting activities** | | |
| 1. | Survey and demarcation including alignment of inspection path of 60 cm width | 0.2 ha/man day | 5 MD/ha |
| 2. | Bush cutting and site clearing | | |
| | Fairly dense shrub growth | 400 – 600 m/day | 20- 25 MD/ha |
| | *Lantana camara* and dense shrub infestation | 200 – 250 m/ day | 40 – 50 MD/ha |
| 3 | Staking and soil working | | |
| | Digging of pits (45cm$^3$) at 3 x 3 m spacing i.e 1111 plants/ha | 15 pits/day | 74 MD/ha |
| | Digging of pits (60 cm$^3$) at 3 x 3 m spacing i.e 1,111 plants/ha | 10 pits/day | 111 MD/ha |
| 4 | Filling of pits (soil + fym 2kg/pit) | 100 pits/day | 11 MD/ha |

| Sr. No. | Operations/Activities | Quantity of work/ day | Total mandays (MD) |
|---|---|---|---|
| 5 | Carriage cost of fym to the planting site (1-2 km) | 160 kg/day | 7 MD/ha |
| 6 | Fencing | | |
| | Bio-fencing (100 x 100 m basis) | 25 m/day | 16 MD/ha |
| | Cattle proof trenching | 6.7 m/day | 60 MD/ha |
| | Barbed wire including preparation of wooden fence post, loading and unloading, digging of fence post holes and other carriage materials | 3 strands – 5 strands | 10 – 14 MD/ha |
| **B** | **Planting activities** | | |
| 7 | Loading and unloading of polythene raised seedling + carrying then to 1 – 2 km distance | 30 seedlings/ person (4 trips/ day) | 10 MD/ha |
| 8 | ETP (naked roots) uprooting, bundling and carrying to 1- 2 km | 100 seedlings/ person (2 trip/day) | 5 MD/ha |
| 9 | ETP tall plants in larger containers with ball of earth | 15 seedlings/ person (4 trips/ day) | 20 MD/ha |
| 10 | Planting and raising of earth around the collar in shape of small mound | 90 – 95 seedlings/person | 11 –1 2 MD/ha |
| **C** | **Post Planting activities** | | |
| 11 | Adding of chemical fertilizers (CAN) @ 20 – 25 g/plant | 550 plants/day | 2MD/ha |
| 12 | Two weeding and hoeing after rainy season i.e Aug. & Sept. | 100/day | 11 MD/ha |
| 13 | Replacement of casuality i.e beating up 10 % of the total plants | 100 seedling/day | 1.1 MD/ha |
| 14 | Making of fire line (Bush cutting + removing of shrubs and unwanted plants upto 5 years) | 10 m length x 5 m with/day/ha (3 times) | 30MD/ha (3 years) |
| | Total | | 446 – 466 MD/ha |

A complete record of all the works done and expenditure incurred in raising the plantation right from the year of planting till its full establishment is essential. This record will be helpful for investment analysis and judging physical achievement made. Performance of the concerned field staff can also be judged to a great extent from the condition of the plantation. The plantation history files should be maintained which contain relevant information on revenue map and ownership, a survey map on a fairly large scale, situation, boundaries, area, year of closure, topography, aspect, geology, soil, hydrologic information including rainfall, frost, hail, wind direction, rights and concession of the villagers, their population and cattle population both domestic and migratory, wildlife population, distance from the nursery. A sample form is given in Annexure 2 and 3 for compiling information of plantation site and operation carried out for maintaining plantation respectively.

## Do it yourself

1. Enlist the promising species which can be used as live fencing in your area.
2. Calculate the cost of fencing for 6 ha land if distance between two fence post is 4 m.
3. Calculate the cost of carrying out of plantation in 5 ha area with nursery raised containerized ETP in your area.

# Literature cited

Anonymous (1978) Forestry for local community development. FAO Forestry Paper 7. FAO forestry department with assistance of SIDA, Rome: 112 p

Anonymous (1983) Simple technologies for charcoal making. FAO Forestry Paper 41. FAO forestry department with assistance of SIDA, Rome: 154 p

Anonymous (2004 a) www.bio.gov.ph International Tree Plantation, Republic of Philippines

Anonymous (2004 b) Wood and its properties as fuel In Single stem tree plantation crops for energy wood production. Forestry Business services (U.K.) Ltd.

Asikainen, A. and P. Pulkkinen (2002) Comminution of logging residues with evolution 910 R chipper, MOHA chipper truck and Morbark 1200 tub grinder. *International J. of Forest Engineering.* Internet edition.

Applegate, G,. Chokkalingam, U and Suyanto. 2001. The underlying causes and impacts of fires in South-east Asia: final report. CIFOR, Indonesia.

Bhardwaj, S.D. and Pankaj Panwar (2000) Impact of mining and rehabilitation of mined areas. Advances in Forestry Research in India. Vol. 22: 151-171.

Bomb calorimetry measurements by Geller at the Indian Institute of Science, Bangalore, India, 1979-80.

Cossalter, Christian and Charlie Pye-Smith (2003) Fast wood forestry- myths and realities. Published by Centre for International Forestry Research, Indonesia.

Evans, J. 1992. Plantation forestry in the tropics. (Second edition) Clarendon Press, Oxford.

Environmental Investigation Agency. 2002. http://www.eiainternational. org/.

FAO. 2000. The global outlook for future wood supply from forest plantations. Working paper GFPOS/WP/03. Rome, Italy.

FAO, 2001 a. Global Forest Resources assessment 2000: main report. FAO Forestry Paper 140. FAO, Rome.

FAO. 2001 b. *State of the World's Forests 2001*. FAO, Rome. http://www.fao.org/

Forestry insights (2004) www.forestryinsights.com

Ghosh, R.C (1977) Handbook on afforestation techniques. ICFRE 313-335.

Hardcastle, P.D. 1999. Plantations: potential and limitations. Draft report for the World Bank Forest Policy Implementation review and Strategy.

Johansson, T.B; Henry Kelly; A.K.N. Reddy and R.H. Williams (1993). Renewable energy : sources for fuels and electricity. Washington : Islands Press.

Kanowski, P.J. (1997). Plantation forestry for the 21st century. In: *Proc. XI World Forestry Congress*. Antalya, Turkey, October 1997. **3**: 23-34. http://sres.anu.edu.au/people/kanowskip.html

Katoch, C.D. (1994) Environmental development cum forest plantation – planning and management. Periodical Expert Book Agency, New Delhi. 131 p

Khanna, L.S. (1993) Principles and Practice of Silviculture. Khanna Bandhu, Dehra Dun.: 472 p.

Klass, D.L. (1990) The U.S biofuels industry. Proceedings of the International Reneable Energy Conference, Honolulu. In : The potential of renewable energy : an interdisciplinary white paper, SERI/TP – 260 – 3674 (U.S. Dept. of Energy) Golden : Solar Energy Research Institute.

Lindenmayer, D.B., Hobbs, R.J. and Salt, D. (2002). Plantation forests and biodiversity conservation. In: *Proc. Australian Forest Plantations 2002*.

Lindenmayer, D.B and J.F. Franklin. 2002. Conserving forest diversity: a comprehensive multiscaled approach. Island Press, Washington D.C

Montalembert, M.R. de and J. Clement (1983) Fuelwood supplies in the developing countries. FAO Forest Paper 42. FAO of United Nations, Rome : 125p.

NCAER. 1985, Domestic Fuel Survey with Special reference to Kerosene, New Delhi: National Council for Applied Economic Research.

Pace University (1990) Environmental cost of electricity. Centre of Environmental leagal studies (ed.), Oceana publication, Inc, New York.

Panwar, Pankaj and S.D. Bhardwaj (2005) Handbook of practical forestry. Agrobios India publication. 210p.

Planning Commission. 1992, Eighth Five Year Plan (1992-97), New Delhi: Planning Commission, Government of India

Prakash, Ram (1986) Forest Management. International Book Distributors, Dehra Dun. 255 p.

Sanghvi, Arun J. (1991) Power shortages in developing countries : impacts and policy implications. Energy Policy 19(5) : 433 – 434.

Saxena, K.G., K.S. Rao, K.K. Sen, R.K. Maikhuri and R.L. Semwal (2001) Integrated natural resource management: approaches and lessons from the Himalayas. Conservation Ecology 5(2): 14

TERI. (1997) Fuel substitution in the rural sector, New Delhi: Tata Energy Research Institute

TERI, (2005) Benefiting the foresters and the community. http://www.teriin.org/teriscope/resupdates/foresters4is.htm

Will, G.M. and Ballard, R. 1976. Radiata pine – soil degrader or improver? New Zealand Journal of Forestry 21(2): 248-252.

World Resource Institute (1994) World Resources. New York-Oxford University Press.

# Annexure- 1

## A. Syllabus of Dr. Y.S. Parmar University of Horticultural and Forestry, Nauni, Solan (Himachal Pradesh)

### SAF –222 Plantation Forestry (2+1)

### Theory

Definition, scope and impediments. Plantation establishment – legal title of land, survey, site selection, site preparation – purposes and methods, Planting-layout, time of planting, planting pattern, spacing, planting methods. Choice of species and intercultural operations. Plantation maintenance – weed control, climber cutting, control of stump growth, staking, singling and pruning.

Nutrition of tree crops – introduction, nutrient deficiencies – causes, occurrence, symptoms, diagnosis and corrective measures. Thinning – definition, objectives. Effect of thinning – physiological and mensurational. Effect of methods of thinning on stand development. Energy and industrial plantation – definition, scope species, establishment, management and impact on environment. Plantation economics.

### Practicals

Study the equipments and materials for establishment of plantation. Site preparation, exercise on planting and tending, study the special technique for difficult sites, exercise on protection of plantation, exercise on plantation planning.

**B. Syllabus as approved by ICAR**

**Plantation Forestry (2+1)**

Introduction, Definition of energy plantations. Energy and economic growth. Classification of energy sources. Wood as a source of energy. Forest as a renewable source of energy. Characteristics of fire wood trees. Efficient use of fire wood (suitable trees for energy planting). Chulahs and wood stoves. Wood energy conservation methods. Non-conventional energy sources. Biogas. Energy savings in industry and houses. Agricultural residues as fuel. Integrated energy planning. Cowdung burning - problem and remedies. Block plantation. Strip plantations. Tree species suitable for different agroclimatic zones.

**Practical**

Study of important fuel wood tree species; different high density energy plantations; Block plantations; energy consumption pattern in rural and urban areas; different chulahs and stoves, design and construction; Astra Chulah and its impact; Biogas plants. Visits to nuclear power static; wind energy station. Energy consumption survey in rural areas among small farmers, marginal farmers, big farmers and agricultural labourers. Determination of calorific value, moisture content and ash content of fuel woods and agricultural residues.

**C. Course content of TNAU**

**FAF $CO_4$ PLANTATION & URBAN FORESTRY (1+1)**

**Theory**

Plantation forestry - Definition - scope - factors - Development - organisation - land and plantation development - land use in forestry. Social and economic factors development - planning. Choice of species - establishment - planting - maintenance - Fertilization.

Urban forestry – History - Distribution and ownership - Benefits –composition of urban forest – environment factors – management – Monitory values of urban forestry.

**Practical**

Study of different types of plantations - Economic consideration in Plantations - Tree parks and Street plantings in cities - Visit to avenue planting at Bhavanisagar and Ecological farm, Bannari – Visit to nearby school and study its urban forest- Visit and study nearby railway line plantings - Visit to botanic garden TNAU-Identification and description of the different landscape components – Designing, layout and drawings for different landscapes.

# Annexure –2

**Format for Plantation work**

| Yr. | Pre-planting activities | | | | Soil working | | | Fencing | | | Seed collection | |
|---|---|---|---|---|---|---|---|---|---|---|---|---|
| | Month | Description of activities | Area (ha) | Man days | Description of activities | Area | Man days | Description of activities | Area | Man days | Quantity | Mandays |
| 1 | | | | | | | | | | | | |
| 2 | | | | | | | | | | | | |
| 3 | | | | | | | | | | | | |

Cont...

| Planting activities | | | | Fertilizer application | | | | Replacement of causality and weeding | | |
|---|---|---|---|---|---|---|---|---|---|---|
| Month | Description of activities | Area (ha) | Man days | Name of fertilizer | Quantity | Area (ha) | Man days | Description of activities | Area | Man days |
| | | | | | | | | | | |

Cont...

| Year | Total mandays | Labour cost (Rs) | Material cost (Rs) | Total expenditure (Rs) | Expenditure/ha or year |
|---|---|---|---|---|---|
| | | | | | |

# Annexure – 3

## Format for maintenance of plantation

| Yr. | Maintenance | | | Tending operations | | | Control burning | | Total man days | Labour rate (Rs) | Total expenditure (Rs) | Expenditure /ha |
|---|---|---|---|---|---|---|---|---|---|---|---|---|
| | Description of activi-ties | Area (ha) | Man days | Description of activities | Area (ha) | Man days | Area (Ha) | Mandays | | | | |
| | | | | | | | | | | | | |